GULF
COAST
DEMISE?

Sponsored by the Harte Research
Institute for Gulf of Mexico Studies,
Texas A&M University–Corpus Christi
Greg Stunz, *General Editor*
John W. Tunnell Jr., *Founding Editor*

GULF
COAST
DEMISE?

CLIMATE CHANGE,
CONSERVATION,
AND SAVING THE
AMERICAN SEA

John B. Anderson

TEXAS A&M UNIVERSITY PRESS | COLLEGE STATION

∞ This paper meets the requirements of ANSI/NISO Z39.48-1992
(Permanence of Paper).
Binding materials have been chosen for durability.
Manufactured in the United States of America.

Library of Congress Control Number: 2024060322
Identifiers: LCCN: 2024060322| ISBN 9781648432798 (printed case) |
ISBN 9781648432804 (ebook)
LC record available at https://lccn.loc.gov/2024060322

A list of titles in this series is available at the end of this book.

CONTENTS

Preface vii

Acknowledgments ix

CHAPTER 1: Geological Evolution of the Gulf Coast *1*

CHAPTER 2: The Role of Sea Level in Coastal Evolution *16*

CHAPTER 3: Climate Change Impacts *52*

CHAPTER 4: Demise of Gulf Coast Deltas *70*

CHAPTER 5: Wave-Dominated Coastal Environments *89*

CHAPTER 6: Estuaries *143*

CHAPTER 7: Demise of Wetlands and Seagrass Meadows *179*

CHAPTER 8: Severe Storms *196*

CHAPTER 9: What Is Being Done? *210*

Glossary 237

References 245

Index 259

PREFACE

Climate change has rapidly forced the Earth into conditions that have not existed historically, leaving us without analogs for current and future change. This book summarizes how different coastal environments responded to sea-level rise and changes in climate over the past several thousand years and how these changes compare to those occurring today. Wetlands are being inundated at alarming rates, with limited opportunity for migration to higher ground. These include the vast Mississippi Delta, western Louisiana Chenier Plain, and the Florida Everglades, which by the year 2050 will have been mostly drowned. Gulf Coast estuaries are also experiencing widespread inundation of their bayhead deltas, which are vital components of estuarine ecosystems. These coastal settings grew significantly during the last 4,000 years. Likewise, barrier islands, peninsulas, and mainland beaches are rapidly eroding following a long history of stability and growth. Hurricanes are getting bigger and wetter and intensifying rapidly before striking the coast, and their impacts are being felt farther inland. This is especially troubling because large portions of the coast are highly vulnerable to storms, given their dense populations and large industrial centers. Decades of ignoring and contributing to these problems have brought us to the point that the demise of many of our most cherished coastal environments is no longer avoidable. Sea-level rise is the main culprit behind coastal change, and it is irreversible, even with large-scale carbon capture. Gulf Coast states have been slow to recognize and deal with coastal change and its causes. Less political debate and more reliance on science are needed to save our coast.

This book is intended for those who desire to become more informed and engaged in saving our coast, including professionals directly involved in coastal conservation, high school and college teachers, and policymakers. While writing this book, I thought about how much scientific detail was needed to present an interesting and compelling discussion on how climate change impacts the Gulf Coast. Too much detail might discourage some readers, while some detail is necessary to demonstrate how and why the coast is changing so rapidly. I have taken the approach of including citations to peer-reviewed scientific papers. I do so because there is too much misinformation on this topic, so it is essential to demonstrate that this book is based on peer-reviewed scientific literature. Peer review is the scientific community's quality control process, and it distinguishes scientific publications from what one might read on social media. I have tried to limit citations to especially relevant ones. I have included many figures in this book to provide documentation for the discussion and to help explain some of the scientific approaches used to understand coastal change. I have also included a glossary of scientific terms used in this book.

ACKNOWLEDGMENTS

This book is based partly on research conducted by my graduate students, who endured the Gulf Coast heat and mosquitoes while collecting cores and geophysical data and inspired me to push ahead. Among them, Dr. Kristy Milliken, Dr. Antonio Rodriguez, Dr. Alex Simms, and Dr. Davin Wallace played significant roles in understanding past and current changes to the Gulf Coast. Dr. David Burkett and two anonymous reviewers provided helpful reviews. I also want to thank Ms. Abagail Chartier and Ms. Laurel Anderton for their highly professional editing assistance. You are fantastic.

I am especially grateful to Dr. Davin Wallace, who provided valuable editing and assistance with figure production. Thank you, Davin, for your support and encouragement and for being there in my time of need.

This book is dedicated to my grandchildren, Kate, Will, and Evie, who spent many days with Doris and me enjoying the coast. You inspired me to write this book. I regret that my generation did not react sooner to preserve our coast for you, your children, and your grandchildren.

GULF COAST DEMISE?

1

Geological Evolution of the Gulf Coast

Climate change has rapidly forced the Earth into conditions that have not existed in historical times, leaving us without analogs for current and future conditions on the planet. Thus, understanding and predicting coastal response to sea-level rise and changing climate is facilitated by understanding how different coastal environments responded to past changes. Background map from Google Earth.

Given its low coastal plain gradients and relatively high subsidence rates, the US Gulf Coast is highly vulnerable to sea-level rise, especially in Texas and Louisiana. It also spans arid to humid climate zones, and paleoclimate records indicate that the region has experienced significant climate variability, especially in Texas. The impacts of sea-level rise and climate change vary for different coastal environments, and the Gulf Coast includes a wide range of coastal settings (fig. 1.1). Thus, predicting future

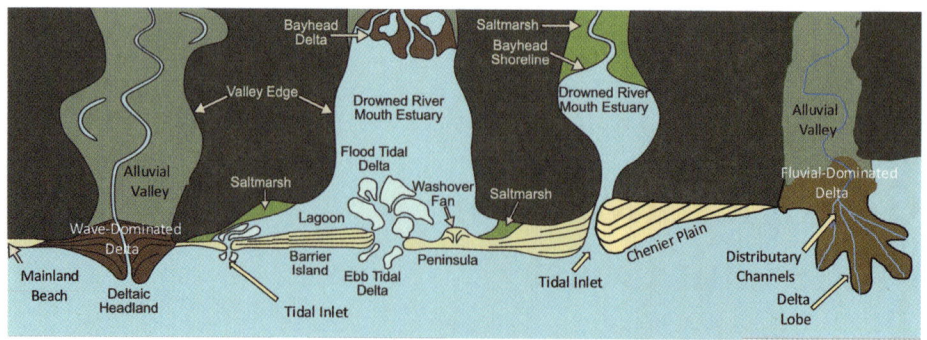

FIGURE. 1.1. The modern Gulf Coast includes many coastal environments that respond differently to rising sea levels, climatic changes, and direct human influences. Modified from Anderson et al. (2022).

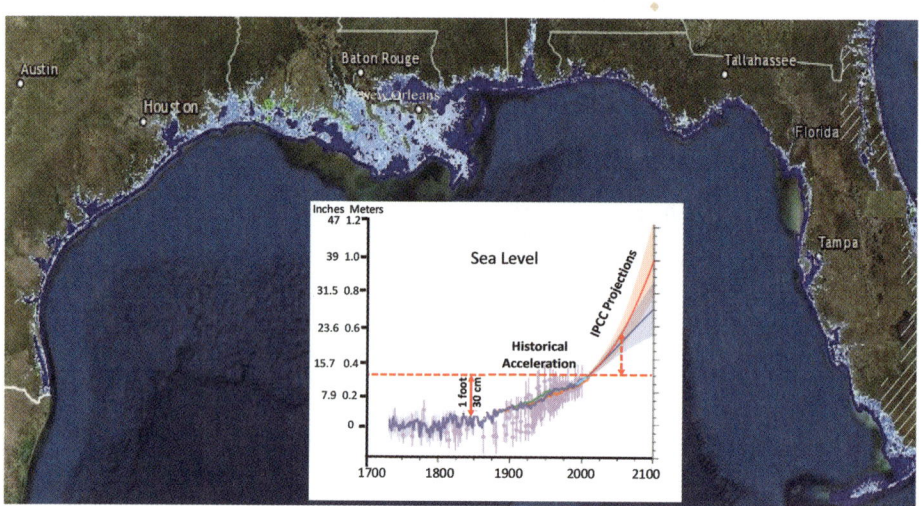

FIGURE 1.2. Model results from the National Oceanographic and Atmospheric Administration (NOAA) show portions of the Gulf Coast that are expected to be drowned by a one-foot (~30 cm) rise in sea level projected for the year 2050, in light blue. Dark blue designates areas currently below sea level. The inset shows the global sea level record from 1700 to 2000, with a one-foot sea-level rise since 1850 when the rate of rise accelerated. The red dashed arrow shows the same magnitude of rise by 2050 based on the latest projections by the Intergovernmental Panel on Climate Change (IPCC). Modified from Anderson et al. (2023).

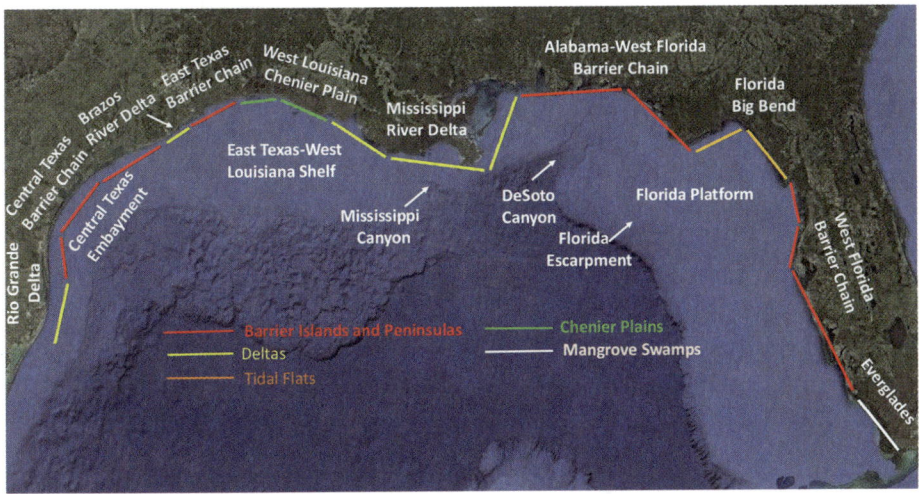

FIGURE 1.3. Map of the northern Gulf of Mexico showing major coastal and bathymetric provinces. Not shown are back-barrier environments, including estuaries, lagoons, and wetlands. Background map from Google Earth.

changes to the coast is more complicated than conveyed by "bathtub" models that drown coastal topography using different rates of sea-level rise. There are also more direct human impacts, such as alteration of riverine sediment supply and dispersal, unbridled urbanization of coastal areas, and alteration of longshore sand transport. These combined influences are taking a heavy toll on the Gulf Coast, and by midcentury, it will be very different from what it is today (fig. 1.2).

As striking as figure 1.2 may be, it fails to convey the full magnitude of coastal change expected in the next few decades. Barrier islands and peninsulas, which span most of the Gulf shoreline, are good examples (fig. 1.3). Their development has varied along the coast. These wave-dominated shorelines respond to multiple factors, so understanding how and why they change is complicated. Knowing how coasts responded to past changes in sea-level rise, sand supply, and other factors provides insight into current changes and should guide efforts to protect our coast.

The Gulf Coast is home to several deltas, including one of the world's largest and most thoroughly studied *fluvial-dominated deltas*, the Mississippi River Delta, and one of the most widely cited examples of a *wave-dominated delta*, the Brazos River delta. The Mississippi River Delta has grown for several thousand years but is now being drowned at an alarming rate.

The once prominent Rio Grande delta met its demise during historical times in response to increasing aridity and excessive water usage. The Colorado and Brazos River deltas have been decreasing in size because of increasing aridity in their drainage basins and human influence on their water and sediment discharge. Several smaller *bayhead deltas* are known to be highly responsive to sea-level rise and climate change. The bayhead deltas in Texas and western Louisiana have changed significantly in response to sea-level rise and changes in climate. At the same time, those in Alabama and Florida have undergone only modest changes. These bayhead deltas will all change significantly by midcentury (see chapter 4).

The Gulf Coast also has some unique coastal settings with highly productive ecosystems. These include Laguna Madre, the western Louisiana Chenier Plain, the Florida Everglades, and Florida's Big Bend area. These coastal settings have had long histories of growth and stability, but they are now changing dramatically. South Louisiana is rapidly losing land because of rising sea levels and human alteration of riverine sediment supply to the coast. The Everglades and Big Bend areas are both being impacted by human alteration of their freshwater supplies, exacerbated by rising sea levels. Thus, coastal change is not a prediction; it is well underway and is expected to intensify over the next few decades.

The myriad modern Gulf Coast environments reflect the region's diverse coastal geology and topography, offshore bathymetry, strong climate gradient, and variable sediment delivery from rivers and offshore sources. To gain an appreciation for how the coast came to include such a broad spectrum of environments, we will begin with a discussion of the last glacial-interglacial cycle, which started ~120,000 years ago (120 ka) when the great ice sheets of the world began to expand, ultimately leading to ~400 feet of global sea-level fall (fig. 1.4a). This was also a period of significantly cooler and wetter conditions.

The Holocene geological epoch spans the last 11,650 years and was the period of Earth's transition into interglacial conditions. It is also the time for which sea-level and climate history are best known, allowing us to better understand how coastal environments responded to variations in rates of sea-level rise and climate change across the Gulf Coast (fig. 1.4b). The Holocene was also the age of human occupation of the coast, which ultimately led to one of the most significant periods of coastal change, the Anthropocene epoch.

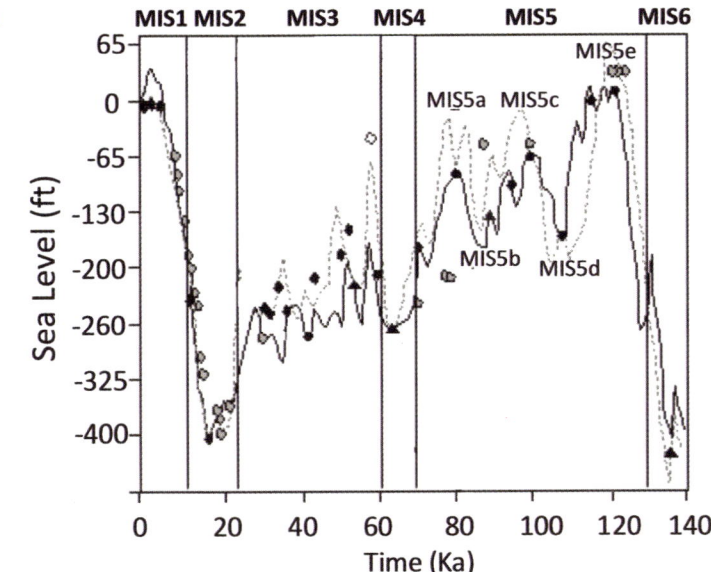

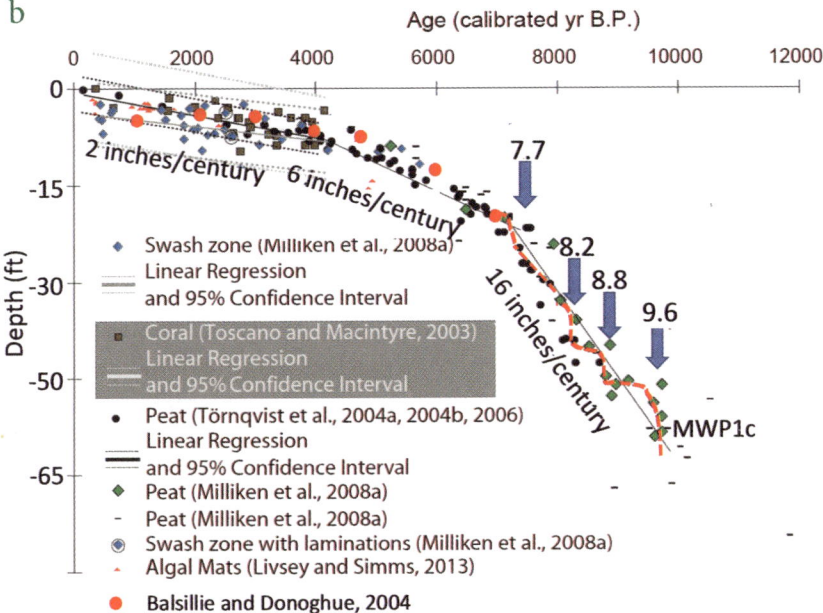

FIGURE 1.4. (a) Global sea level record for the past 140,000 years based on oxygen isotope records and actual sea-level markers (black and gray dots). Marine Isotope Stages (MIS1–6) are used by convention to distinguish various stages of rise and fall. (b) Gulf of Mexico sea-level record for the last 10,000 years with sources indicated. Blue arrows designate episodes of punctuated sea-level rise. Entire figure modified from Anderson et al. (2022).

The modern Gulf Coast is divided into eastern and western sectors, with different physiography, subsidence histories, and sedimentation. DeSoto Canyon divides these two sectors (fig. 1.3). West of the canyon is the Gulf of Mexico sedimentary basin, which contains up to 40,000 feet of sedimentary deposits that have accumulated since Triassic time. The great thickness of sediments in the basin reflects the role of large rivers, specifically the Mississippi, Rio Grande, Brazos, and Colorado, in delivering sediment to the Gulf of Mexico for millions of years. The load of this thick sediment column helps drive subsidence along the western Gulf Coast.

The eastern sector of the Gulf Coast includes the Alabama and Florida coasts, which are bounded offshore by the 160-mile-wide, low-gradient Florida Platform (fig. 1.3). Cretaceous strata, mostly carbonates, lie within 6,500 feet of the seafloor and reflect slow sediment accumulation relative to the Gulf of Mexico Basin. This thinner sediment column has resulted in slower subsidence and a virtual absence of salt-related faulting, which can significantly influence coastal subsidence. Riverine sediment contributions to the Florida and Alabama coast are small relative to those of the western Gulf, with the Apalachicola and Alabama-Tombigbee Rivers being the largest sediment sources. As a result, carbonates, which form in the region's relatively warm, clear waters, are a significant component of the sediment column. The Florida and Alabama continental shelves tend to be sand rich, without significant river-supplied silt and clay. In contrast, thick mud drapes the Louisiana and Texas shelves, and their coasts are composed mainly of fine sand derived from the Mississippi River and the reworking of offshore deltas. The product is white sandy beaches and clear emerald-green waters along the west Florida and Alabama coast versus the darker-colored beaches and turbid coastal waters of Louisiana and Texas.

The Late Pleistocene: Setting the Stage for Gulf Coast Evolution

The modern coastal plain topography and offshore bathymetry of the Gulf Coast formed during the past ~120,000 years as sea level fluctuated hundreds of feet in response to expansion and decay of ice sheets. For our discussion, we will use Marine Isotope Stages (MIS) to designate time intervals marked by major sea-level events, as shown in figure 1.4a.

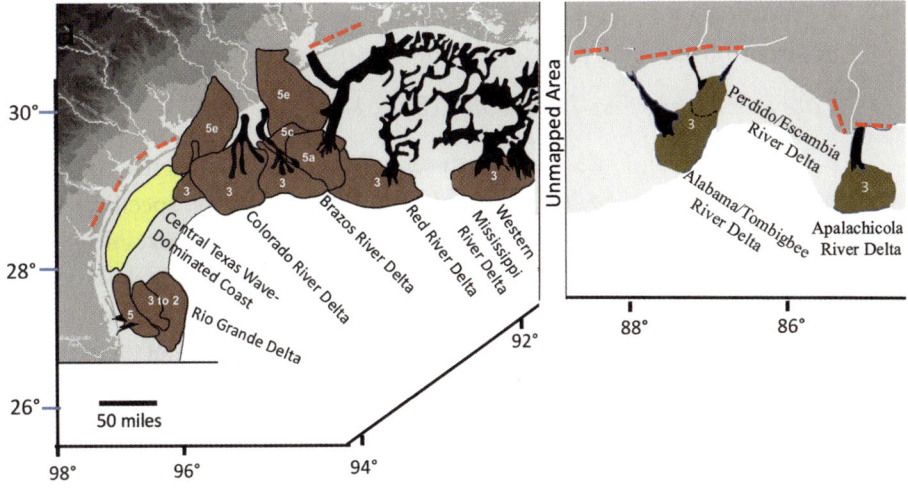

FIGURE 1.5. Offshore deltas (brown) of the western Louisiana–Texas (left) and northwest Florida (right) continental shelves. The dashed red line is the Ingleside Paleoshoreline. The age of deltas is designated using oxygen isotope stages 5e through 3 (see fig. 1.4a). Note that the Brazos, Colorado, and Rio Grande deltas decrease in age in an offshore direction, indicating seaward growth as sea level fell. The ages of Texas deltas are from Anderson et al. (2004), and the ages of offshore Florida deltas are from Bart and Anderson (2004) and McKeown et al. (2004). Light gray designates the continental shelf, and black designates former river channels. Figure modified from Anderson et al. (2022).

The Last Interglacial (MIS 5e)

The last interglacial, which peaked around 120 ka, provides a model for the Gulf Coast's interglacial setting. Sea level at that time was about 15 to 25 feet higher than at present, and the Gulf Coast shoreline was situated between 18 and 30 miles landward of the modern shoreline in east Texas and Louisiana but closer to the current coastline along the higher-gradient Mississippi, Alabama, and Florida coast. This old shoreline remains as a distinct sand ridge along portions of the Gulf Coast, known as the *Ingleside Paleoshoreline* (fig. 1.5). Characterized by remnants of old beaches composed of white sandy soils, it is bounded to the south by a relatively flat, marshy coastal plain and to the north by greater relief associated with exposures of Pleistocene and older strata. As we will see, it is a prominent physiographic feature that defines the northern shoreline of numerous coastal barriers, estuaries, lagoons, and wetlands. It is also a topographic barrier to their landward migration.

Late Pleistocene Sea-Level Fall (MIS 5d–MIS 3)

Between MIS 5 and MIS 3, ice sheet expansion in both hemispheres resulted in approximately 400 feet of sea level fall (fig 1.4a). The overall fall was punctuated by several sea-level oscillations with magnitudes of 15 to 50 feet, each lasting ~20,000 years. These oscillations record episodes of ice sheet growth and decay caused by astronomically driven climate cycles (chapter 2). Differences in the exact duration and magnitude of these events were modulated by oceanic and atmospheric circulation effects and by the variable response of the ice sheets to atmospheric and oceanographic influences.

The response of coastal environments to sea-level oscillations during the overall MIS 5–3 fall was one of seaward migration of the coast that began at the Ingleside shoreline. The coast differed from that of today, with extensive deltas and cypress swamps stretching across Louisiana and east Texas (fig. 1.5). Large deltas also existed offshore of west Florida. They were formed by the ancestral Alabama-Tombigbee, Perdido-Escambia, and Apalachicola Rivers and remain evident today as bathymetric highs on the outer continental shelf. These deltas indicate high sediment discharge to the basin because of humid climate conditions, downcutting of river valleys, and expansion of their drainage basins as sea level fell. A lack of river sediment input to the central Texas coast resulted in an extensive wave-dominated shoreline in that area.

The Florida Platform was shaped more by wave erosion than by river incision and delta formation. This explains the relatively flat continental shelf relative to offshore Louisiana and Texas. It also reflects the slow subsidence rate on the Florida continental shelf, which has resulted in the erosion of younger strata. One exception is an exposure of isolated swamp deposits approximately 18 miles offshore of Gulf Shores, Alabama. Cypress stumps from the area have yielded ages between 73 and 56 ka (DeLong et al., 2021). Otherwise, Pleistocene deposits are restricted mostly to the outer continental shelf, where faster subsidence rates occur.

The Last Lowstand (MIS 2)

During the MIS 2 sea-level lowstand, which peaked at ~18 ka, sea level was about 400 feet below the present level (fig. 1.4a). During that time, the shoreline was located at the edge of the continental shelf, and rivers and streams extended their valleys across the shelf. The irregular surface cut by these rivers is easily recognized in seismic profiles. It has been mapped in relative detail on the western Louisiana and Texas continental shelves and the Mississippi continental shelf (fig. 1.6).

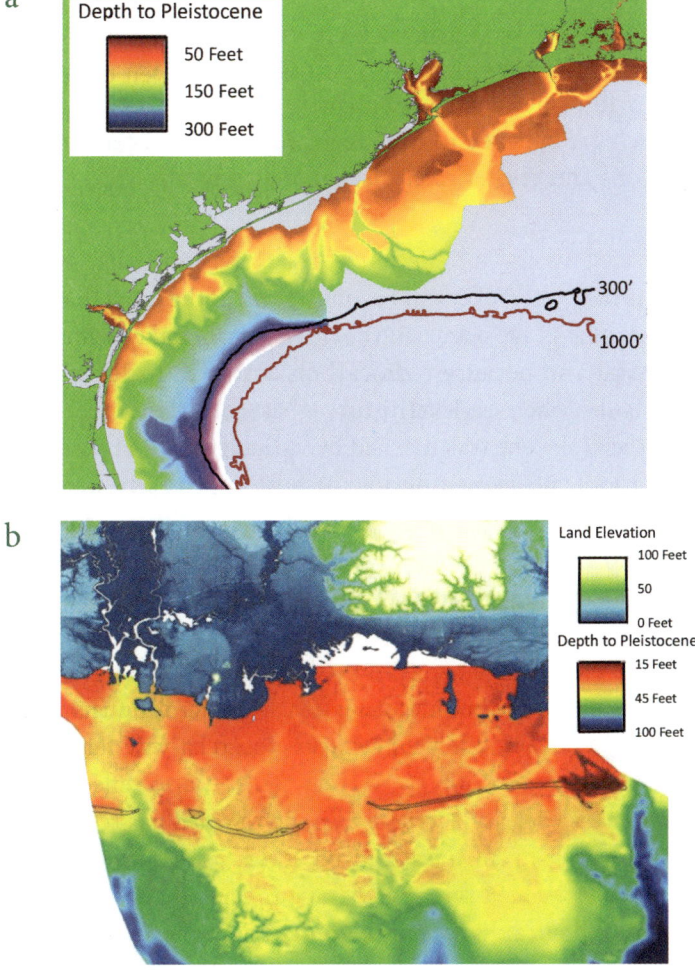

FIGURE 1.6. *Facing*, Maps showing Pleistocene relief on the (a) Texas and western Louisiana continental shelf (modified from Simms et al., 2007b) and (b) Mississippi Sound and inner continental shelf (modified from Hollis et al., 2019). This relief was formed during the last lowstand in sea level (MIS 2, fig. 1.4a). During that time, the continental shelf was subaerially exposed, and rivers incised and extended their valleys (shown in yellow) across the continental shelf.

The Postglacial Transgression (MIS 1)

The initial deglaciation phase occurred between ~18 ka and 12 ka, resulting in a sea-level rise of ~200 feet (fig. 1.4a). During this time, shorelines
shifted rapidly landward, leaving little in the way of a geological record
of coastal evolution. Exceptions to this were the offshore deltas formed
by the Rio Grande and the Mississippi, Brazos, and Colorado Rivers.
These rivers continued to construct large deltas during the early stages
of the transgression despite the relatively rapid rate of sea-level rise. The
continued growth of these deltas implies high sediment delivery rates to
the coast, which are believed to have resulted from persistent cooler and
wetter conditions and greater river discharge during this time.

The Holocene

The *Holocene* is the period of coastal evolution we understand best because
coastal deposits of this age occur on the inner continental shelf and along
the modern coast and because radiocarbon dating provides reliable age
constraints. It is also when sea-level history is best constrained (fig. 1.4b). The
beginning of the Holocene was marked by variable sea-level rise, with the
average rate periodically exceeding 20.0 inches/century. This was followed
by a decrease in ice sheet decay and a rise in sea level that proceeded in
steps. From ~10.0 ka to 7.5 ka, the average rate of rise was 16 inches/century
(4.2 mm/yr). Between 7.5 ka and 4.0 ka, the rate was 6 inches/century (1.4
mm/yr); after 4.0 ka, the rate decreased to about 2.0 inches/century (0.5
mm/yr). For the sake of discussion in this and the following chapters, we
will use these changes in the rate of sea-level rise to designate the early
Holocene (10.0 to 7.5 ka), middle Holocene (7.5 to 4.0 ka), and late Holocene
(4.0 ka to the present) (fig. 1.4b).

As the rate of sea-level rise decreased, so did the rate of landward shoreline migration, resulting in increased preservation of coastal deposits.
These deposits have yielded the most detailed record of coastal evolution
for any time in geological history. The average rate of sea-level rise in the
Gulf of Mexico during the past three decades was 12 inches/century, so the
early Holocene provides the best analog for coastal response to this rate
of rise. However, it is also important to remember that the early Holocene
climate was wetter than the climate today, so sediment supply to the coast
was relatively high. By middle Holocene time, more arid conditions in
the Brazos, Colorado, and Rio Grande drainage basins resulted in lower

sediment supply from these rivers, so they could no longer nourish offshore deltas. Because of its much greater sediment supply to the coast, the Mississippi River continued to construct extensive delta lobes on the continental shelf throughout the late Holocene. This long history of delta growth is now ending (chapter 4).

Across the Gulf Coast, rivers with smaller drainage basins and lower sediment discharge could not fill their valleys with sediment and construct offshore deltas. Instead, their valleys were flooded during sea-level rise to form estuaries. Meanwhile, sand derived from the erosion of offshore deltas nourished coastal barriers, chenier plains, and *tidal deltas*. Sand eroded from the Apalachicola, Perdido-Escambia, and Mobile deltas was spread across the continental shelf as an extensive sand deposit known as the *MAFLA* (Mississippi, Alabama, and Florida) *Sheet Sand*. This sand layer is still exposed at the seafloor from the Apalachicola River mouth to Dauphin Island, Alabama, and was the primary source of sand for the formation of the mainland beaches and coastal barriers of northwest Florida and Alabama (fig. 1.7). The barrier islands of Mississippi were nourished mainly by sand eroded from the St. Bernard lobe of the Mississippi Delta and the Biloxi and Pascagoula deltas. In Texas, sand that nourished the coast was derived primarily from the offshore Mississippi, Brazos, Colorado, and Rio Grande deltas (fig. 1.5).

The Early Holocene: Coastal Response to Rapid Sea-Level Rise

The average rate of sea-level rise during the early Holocene was 16 inches/century (fig. 1.4b), which is in the range of sea-level rise predicted for the next few decades. Furthermore, the rise in sea level during the early Holocene was punctuated by episodes when the rate was at least twice the average. It is generally believed that these pulses of rapid rise were caused by the collapse of major ice streams, probably in Antarctica. Thus, the early Holocene provides a glimpse of coastal response to conditions that would occur if large West Antarctic *ice streams*, like the Pine Island and Thwaites ice streams, were to collapse (chapter 2).

Unfortunately, Gulf coastal evolution during the early Holocene remains unclear because of a scarcity of exposures of this age and the limited number of offshore studies. The low number of exposures is mainly the result of erosion during sea-level rise and landward translation of the shoreline, known as *transgression*. During transgression, storm waves erode the seafloor at water depths shallower than 25 to 30 feet, removing the upper part

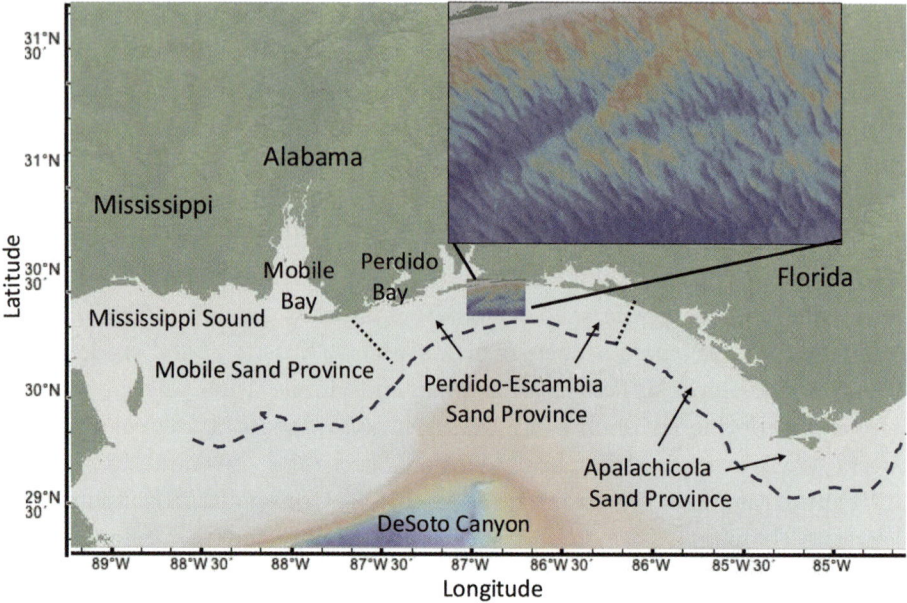

Figure 1.7. Map of the Mississippi, Alabama, and west Florida continental shelf showing the extent of the MAFLA Sheet Sand (dashed line), which is divided into the Mobile, Perdido-Escambia, and Apalachicola sand provinces based on their primary fluvial sources (fig. 1.5). Inset image is a detailed bathymetry map of the inner continental shelf near Pensacola, Florida, showing large ridges associated with the MAFLA Sheet Sand. From GeoMapApp.

of the sediment column. Geologists refer to this process as *transgressive ravinement* because of the widespread extent of the erosion. Early Holocene sediments have been eroded and reworked primarily in Florida, Alabama, and Mississippi to form the MAFLA Sheet Sand. The continental shelf is blanketed by Mississippi River Delta deposits in Louisiana and much of Mississippi. These deposits have yielded a delta evolution record extending back approximately 4,600 years. Still, little is known about how the delta responded to the more rapid sea-level rise of the early Holocene, leaving us without analogs for the current changes to the delta.

In central and south Texas, the stratigraphic record of coastal evolution during the early Holocene is primarily masked by a thick layer of mud known as the *Texas Mud Blanket* (Weight et al., 2011). As a result, the east Texas continental shelf is the only area of the Gulf where early Holocene coastal deposits have been sampled and studied in detail (Anderson et al.,

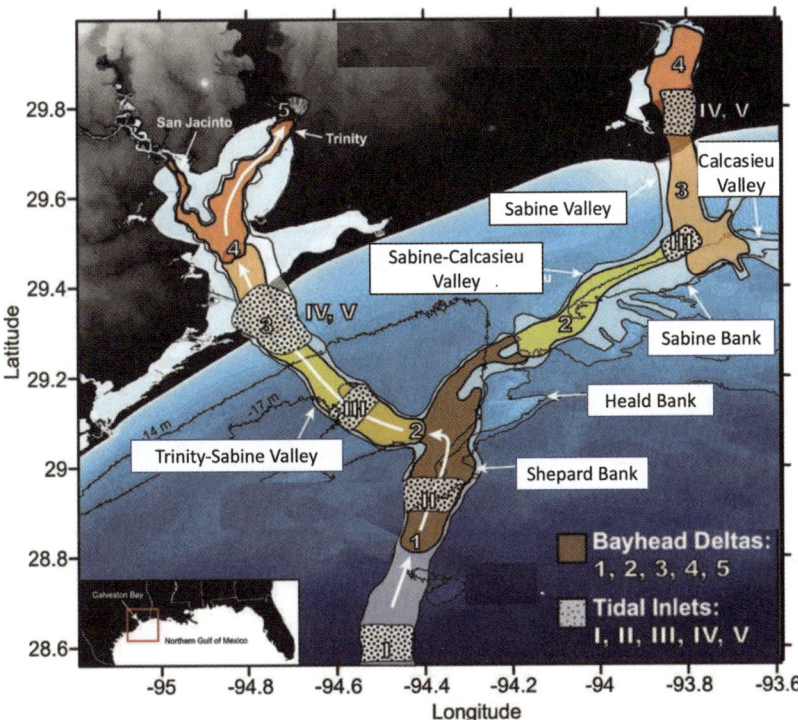

FIGURE 1.8. Paleogeographic map of the east Texas continental shelf showing portions of offshore Trinity River and Sabine River incised valleys and the bayhead deltas and tidal inlets formed within these valleys during the late Pleistocene and early Holocene. Tidal inlet deposits show an overall pattern of landward-stepping shoreline locations. Also shown are offshore Sabine, Heald, and Shepard Banks, former coastal barriers drowned in place during sea-level rise. These landward-stepping coastal environments record punctuated sea-level rise during the early Holocene (fig. 1.4b). Modified from Anderson et al. (2022).

2022). These deposits occur within offshore river valleys and include estuarine and coastal sediments deposited below the depth of wave erosion (fig. 1.8). There are also sand banks on the shelf, including Sabine Bank, Heald Bank, and Shepard Bank. Seismic data and sediment cores collected from these banks have yielded compelling evidence that they are former coastal barriers that drowned in place and were later reworked during sea-level rise (Rodriguez et al., 1999). The ages and spacing of these banks indicate that the shoreline shifted landward at rates as high as 20 feet/year in response to episodes of punctuated sea-level rise.

The Evolution of the Modern Coast

By late Holocene time, the rate of sea-level rise had decreased significantly, setting the stage for modern coastal environments to develop. It was also a time of increased human occupation of the coast (Worrall, 2021). The slower rate of sea-level rise was conducive to the preservation of coastal deposits, but the stratigraphic record of coastal evolution is incomplete in most areas. Holocene coastal deposits have been eroded and buried beneath the MAFLA Sheet Sand offshore of Florida and Alabama. Hence, the record of coastal evolution is limited mainly to onshore and nearshore areas. Mississippi has a more complete stratigraphic record that has been the focus of research by Dr. Ervin Otvos and Dr. Davin Wallace at the University of Southern Mississippi. In Louisiana, Holocene strata are pretty thick and have yielded a record of the Mississippi River Delta's evolution. The Texas continental shelf has provided the best record of how different coastal environments responded to sea-level rise and climate change during the Holocene (fig. 1.9).

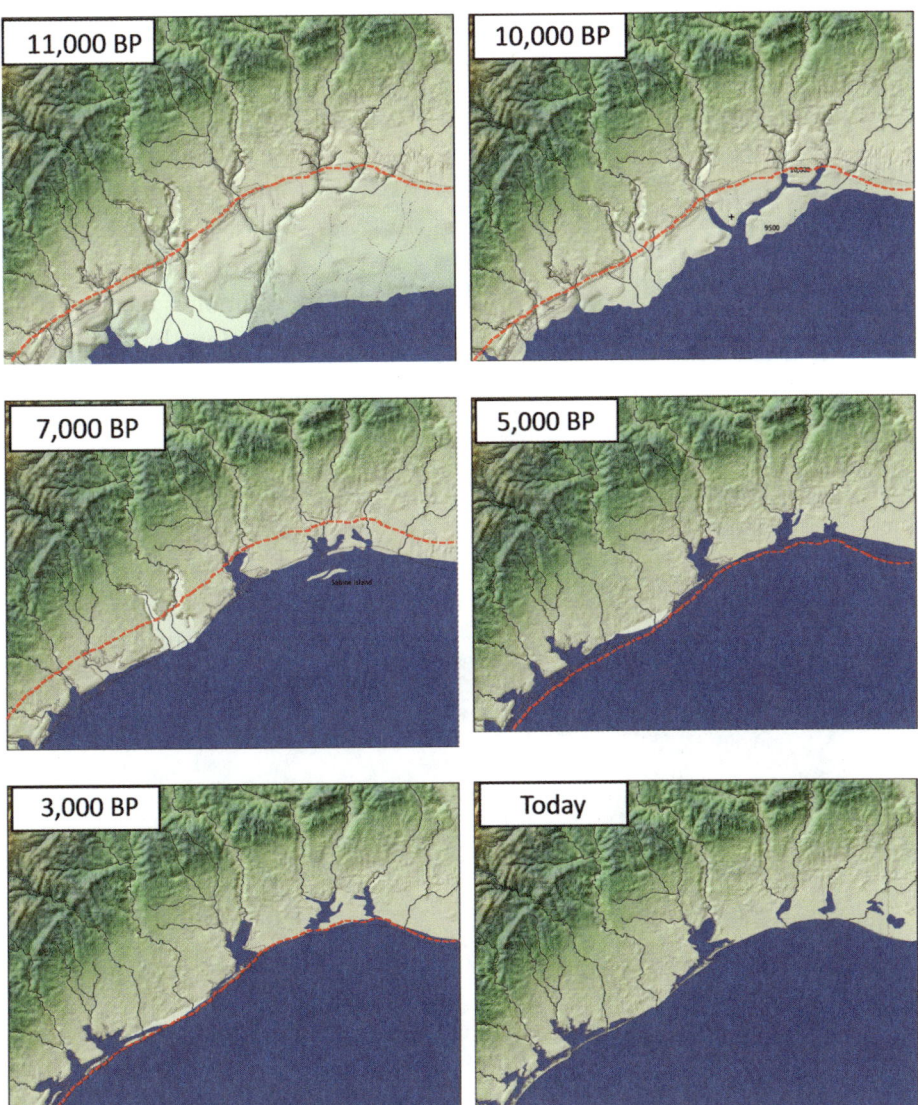

FIGURE 1.9. Paleogeographic maps constructed from seismic data illustrate the Holocene evolution of the west Louisiana and east Texas coast. The red dashed line is the modern shoreline. Modified from Worrall (2021).

2

The Role of Sea Level in Coastal Evolution

Climate change deniers like to say that the climate has always changed, so what's the big deal? The problem is the unprecedented rate at which climate change and its impacts are occurring and the fact that humans are exacerbating these impacts.

For example, the current rate of sea-level rise in the Gulf of Mexico has increased sixfold in the past century. It has been over 7,500 years since the rate was as fast as it is today, and we are still unprepared to accept the consequences.

The rate of sea-level rise is accelerating rapidly. The result is significant alteration and loss of coastal environments and ecosystems and increased vulnerability to storms, coastal flooding, and saltwater intrusion into coastal aquifers and rivers. These impacts vary regionally because the rate of rise varies along the coast with different rates of land subsidence, which in parts of coastal Louisiana exceed the rate of eustatic rise (*eustasy*). The combined eustatic rise and subsidence is called *relative sea-level rise* (RSLR). This chapter provides background on the factors that influence sea level and how sea-level changes are measured over different timescales. One of the most significant uncertainties in predicting future sea-level rise has to do with rates of greenhouse gas emissions in coming decades. This is manageable, but progress to date has been slow. The other is the contribution from glaciers and ice sheets, which could double the rise rate over the next several decades. This contribution is irreversible over decadal timescales.

Glacial Cycles

One question I commonly hear is, "How do we know that the rise in sea level is not simply part of the natural cycle of glacial and interglacial conditions that have occurred for millions of years?" It is a good question.

Throughout geological time, the Earth has shifted from periods of glaciation (*icehouse conditions*) to interglacial periods (*greenhouse conditions*) that occurred over geological timescales. Large changes in Earth's climate were a product of the movement of continents in and out of polar latitudes, driven by plate tectonics, resulting in sea-level rise and fall of tens to hundreds of feet over geological timescales. Among the more spectacular records of these changes are alternations between marine and continental rocks in sedimentary basins and on continental shelves.

During the late Paleozoic, the Southern Hemisphere megacontinent Gondwana began to break apart, with Antarctica drifting south. At the same time, other continents remained near or north of their previous locations. By late Cretaceous time, Antarctica was near its current polar location. Despite this southerly drift, the continent remained relatively warm and covered by forests because of unusually warm conditions during the *Paleocene-Eocene Thermal Maximum* (PETM). The Northern Hemisphere was also warmer, and subpolar regions were occupied by plants and animals that migrated from warmer latitudes. It was not until about 35 million years ago that the Antarctic Ice Sheet formed during the final isolation of Antarctica from other Gondwana continents and the end of the PETM. The continent's isolation set the stage for circum-Antarctic atmospheric and oceanic circulation, resulting in a polar climate regime. It was not until approximately 3.5 million years ago that the Northern Hemisphere continents in subpolar latitudes began to be cold enough to support ice sheets.

The current "icehouse" state of the Earth is unique in that it is a bipolar glaciation, with ice sheets in both hemispheres, in contrast to previous glacial periods when ice sheets were mainly centered over megacontinents. During the *Last Glacial Maximum* (LGM), which peaked approximately 20 ka, enough ice was stored in the ice sheets to lower global sea levels about 400 feet below present levels (fig. 1.4a). Most of that eustatic change was caused by the expansion of Northern Hemisphere ice sheets, which, because of their relatively low-latitude locations, were more sensitive to changes in atmospheric temperatures than the Antarctic Ice Sheet (fig. 2.1).

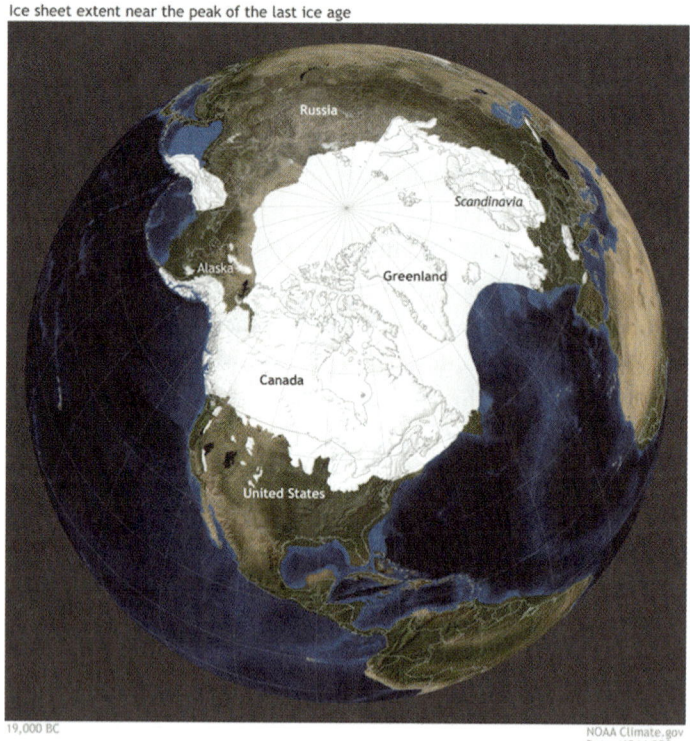

FIGURE 2.1. Maximum extent of Northern Hemisphere ice sheets during the Last Glacial Maximum. From NOAA Climate.gov.

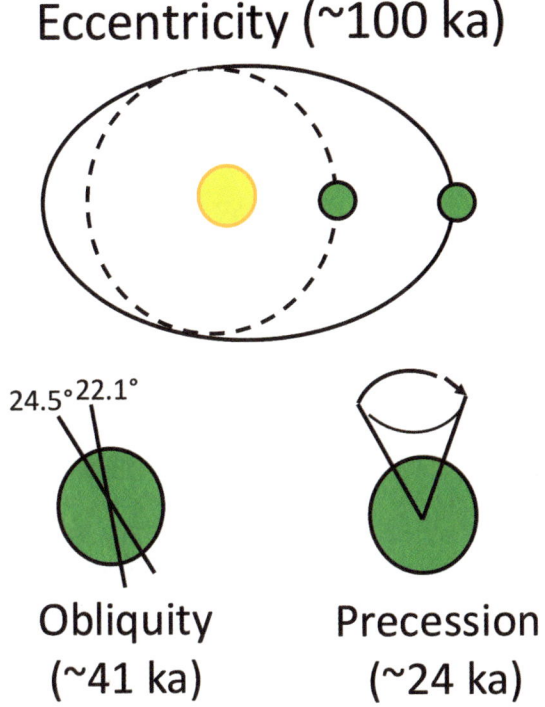

FIGURE 2.2. Milankovitch's astronomical cycles are the main drivers of climate change at geological timescales.

As a result, Northern Hemisphere ice sheets have had a history of advance and retreat that is strongly correlated to natural climate cycles controlled by astronomical changes. These are known as *Milankovitch cycles*, and they are driven by variations in the Earth's rotation around its axis and its revolution around the sun (fig. 2.2).

The Earth's orbit around the sun varies from nearly circular to eccentric, resulting in changes in its spacing from the sun every 100,000 years. This is known as the 100,000-year *eccentricity* cycle. The second astronomical driver of Earth's climate is the tilt of the Earth relative to its axial plane, which varies between 22.1° and 24.5°. This results in variations in the amount of solar radiation reaching the poles versus the equator, with increased tilt resulting in greater heating at the poles. This is known as the *obliquity* cycle, which has a period of about 41,000 years. The third astronomical influence on global climate results from changes in the Earth's axis of rotation caused by tidal forces exerted by the sun and the moon on the Earth. This is known as *axial precession* and has a frequency between 19,000 and 24,000 years. These astronomically driven climate cycles are now known to have dominated past glacial-interglacial cycles, and therefore sea-level rise and fall.

Measuring Changes in Ice Volume

Changes in the volume of ice on the continents versus the volume of water in the oceans result in changes in oxygen isotopes (O^{18} and O^{16}) within the ocean (fig. 2.3). The buildup of ice sheets removes water from the oceans, resulting in lower sea level and ocean water that is enriched in O^{18} relative to O^{16}. The O^{18} in water that evaporates from the oceans is transported in rain clouds onto the continents, where it is preferentially removed by precipitation at lower altitudes. In comparison, snow and ice formed at higher altitudes are enriched in O^{16}. Scientists use mass spectrometers to measure the relative proportions of oxygen isotopes in *foraminifera* shells from sediment cores and air bubbles in ice cores to obtain records of changes in global ice volume. The result is a record of ice volume that extends back hundreds of thousands of years in ice cores and millions

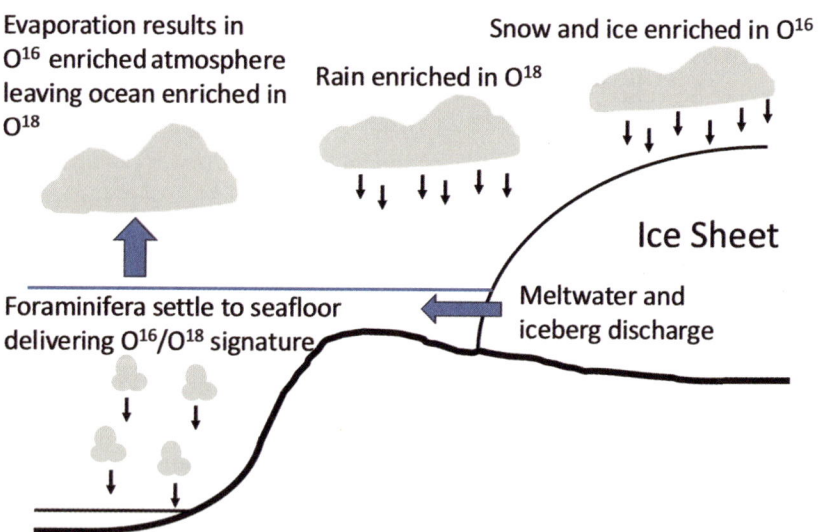

FIGURE 2.3. Changes in the O^{18}/O^{16} concentration of the ocean and ice sheets.

of years in sediment cores. These ice volume records are then converted to sea-level records. Isotopic data can also be used to measure oceanic and atmospheric temperatures and greenhouse gas concentrations over geological time, allowing scientists to investigate the natural causes of climate change and associated eustatic change.

One of the best ice core records comes from Lake Vostok, Antarctica. It spans four 100,000-year glacial-interglacial cycles, providing a record of temperature, sea level, and carbon dioxide concentrations over that time interval (fig. 2.4). The temperature and CO_2 records reveal a close linkage between climate cycles and CO_2 concentrations. Ironically, this connection has been used by critics of climate change to argue that CO_2 has varied naturally through time and that the current increase in CO_2 concentrations is part of this natural cycle. Atmospheric CO_2 concentrations and global temperatures have indeed varied together, which is the expected greenhouse effect. However, the most compelling observation is that the historical increase in CO_2 is unprecedented for the last four Milankovitch-driven glacial-interglacial cycles. Studies have shown that this historical increase in CO_2 can be explained only as a result of the combustion of fossil fuels.

As shown in figure 2.4, the 120,000-year astronomical cycles of global climate change have resulted in eustatic cycles with amplitudes of about 400 feet. This magnitude of change could have been caused only by significant changes in the volumes of ice sheets. Geologists have long thought these changes were driven by oscillations in Northern Hemisphere ice sheets and that the Antarctic Ice Sheet was relatively immune to these astronomical climate influences. However, recent decades of research have revealed that the Antarctic Ice Sheet has also experienced significant volume changes that were generally in phase with Northern Hemisphere changes.

So, where are we currently in terms of the Earth's natural climate cycle? Are we still experiencing global warming and rising sea levels because the Earth is transitioning out of a glacial episode? Recognizing that the global sea level is below its level during the last interglacial ~120 ka is relevant. At that time, global sea level was between 18 and 30 feet higher than the present level. In the Gulf Coast region, this sea-level highstand is prominently marked by the Ingleside Paleoshoreline, which is located well inland of the modern coast at an elevation of 18 to 24 feet (fig. 1.5). This suggests that the Antarctic and Greenland Ice Sheets were smaller than they are now and that the current ice sheet configuration is still not in a true interglacial

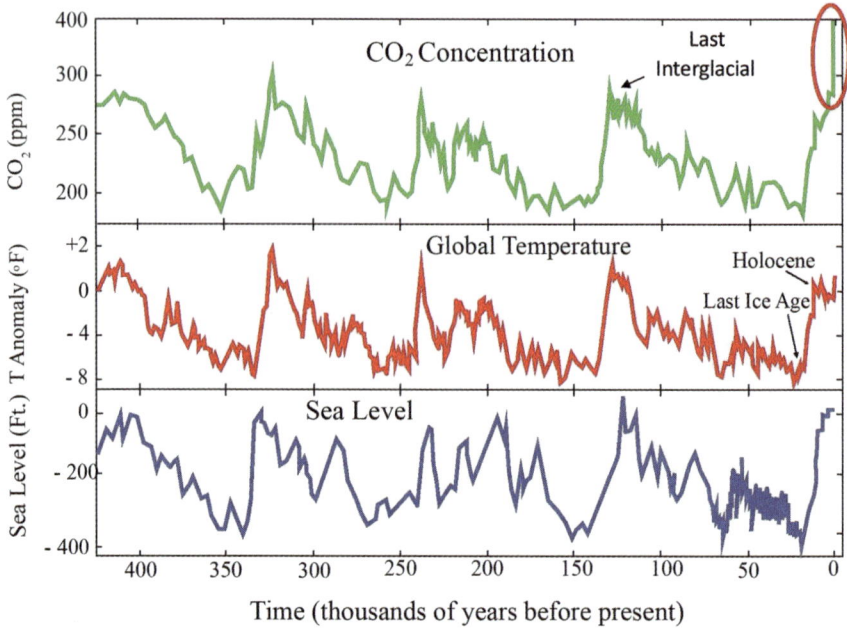

FIGURE 2.4. Oxygen isotope ratios recorded in air bubbles in ice cores from Vostok Station, Antarctica, provide a long-term record of atmospheric temperatures and ice volume, which is used to measure sea-level change. The air bubbles are also used to measure atmospheric CO_2 concentrations through time. These independent proxies for climate change span more than 400,000 years. Note the strong correlation between CO_2, atmospheric temperature, and sea level. Note also that current CO_2 concentrations are significantly higher than at any other point during this time. Also, this long-term record does not capture the increases in atmospheric temperature and sea-level rise in recent decades, which are measured using other methods. Red circle shows the historical increase in CO_2. Modified from John Englander's website, https://john-englander.net.

mode. The problem is that humans have greatly accelerated the rate of climate change and sea-level rise.

As we will discuss, there is compelling evidence that ice sheets are currently decreasing in size and pose the greatest threat as far as future sea-level rise. If the past is a key to the future, sea level along the Gulf Coast could rise another 18 to 24 feet before the maximum interglacial state is reached. However, oxygen isotope records indicate that we have already exceeded previous interglacial conditions, so the magnitude of sea-level rise could be even higher. The concerns are the rapid rate at which glaciers and ice sheets are approaching interglacial conditions and how much sea level will rise given the unprecedented greenhouse gas concentrations shown in figure 2.4. To better understand the role of ice sheets in past and future sea-level rise, we first need to discuss the factors that influence ice sheet stability and their contribution to sea-level rise.

Sea-Level Rise since the Last Glacial Maximum

During the Last Glacial Maximum (LGM), which occurred ~22 ka to ~18 ka, large ice sheets existed over Canada and the northern United States, most of northern Europe and the British Isles, Scandinavia, northern Asia, Russia, and Greenland (fig. 2.1). In the Southern Hemisphere, glacial expansions occurred in New Zealand and South America. In Antarctica, the ice sheet was thicker than it is today and advanced far out onto the continental shelf. The volume of water transferred from the oceans to the continents accounts for most of the ~400 feet of sea-level lowstand of the LGM.

Around 18 ka, Earth's climate shifted out of the "icehouse" state, and ice sheets began to retreat. The record of sea-level rise following the LGM is recorded in coral reefs around the globe. These records indicate that sea level rose approximately 200 feet between 18 ka and 10 ka (fig. 1.4a). The overall rise was episodic, with periods when the rate was nearly an inch per year. These episodes of rapid rise are known as meltwater events. The early sea-level rise following the LGM was driven mainly by deglaciation of Northern Hemisphere ice sheets, with the Antarctic Ice Sheet playing an increasing role after ~14 ka. The rate of rise decreased after ~12 ka. This is roughly the period known as the Holocene epoch (11,650 ka to the present), which is the postglacial period. As the rate of sea-level rise decreased, the preservation of coastal deposits on continental shelves increased, providing more opportunities to compile better sea-level records (fig. 1.4b). The more accurate sea-level curves allow research to compare rise rates to the geological record of coastal change.

Between 10 ka and 7.5 ka, the average rate of sea-level rise in the Gulf of Mexico was 16 inches/century (4.2 mm/yr) (fig. 1.4b). Antarctica had assumed a significant role in rising sea levels by this time. The sea-level curve in figure 1.4b shows four episodes of punctuated sea-level rise during the early Holocene, around 9.6, 8.8, 8.2, and 7.7 ka. The magnitude of these events averaged about 5 feet within a few centuries. Given the rate and magnitude of these events, they are believed to record episodes of rapid ice sheet retreat and the release of water from large glacial lakes known to have stored vast quantities of glacial meltwater. This water was suddenly released into the oceans when the natural dams for these lakes ruptured.

As we will see, these episodes of rapid rise had a dramatic impact on coastal environments. Indeed, they dominated coastal evolution during the early Holocene.

By ~7.5 ka, the rate of ice sheet decay in Antarctica had slowed significantly, and the average rate of sea-level rise decreased to 6 inches/century (1.5 mm/yr). By ~4.0 ka, the average rate fell to 2 inches/century (0.5 mm/yr), which marks the end of significant ice sheet contribution to sea-level rise before historical time.

The Historical Acceleration of Sea-Level Rise

Until recently, evidence for the historical acceleration of sea-level rise came mainly from tide-gauge records, and this evidence was widely debated. This debate was fueled by the relatively short duration of most tide-gauge records and the different statistical approaches used to analyze these data. However, satellite data acquired since 1993 have provided indisputable evidence that the rate is accelerating (fig. 2.5). It is now generally accepted that global sea-level rise (eustasy) has steadily increased since 1900.

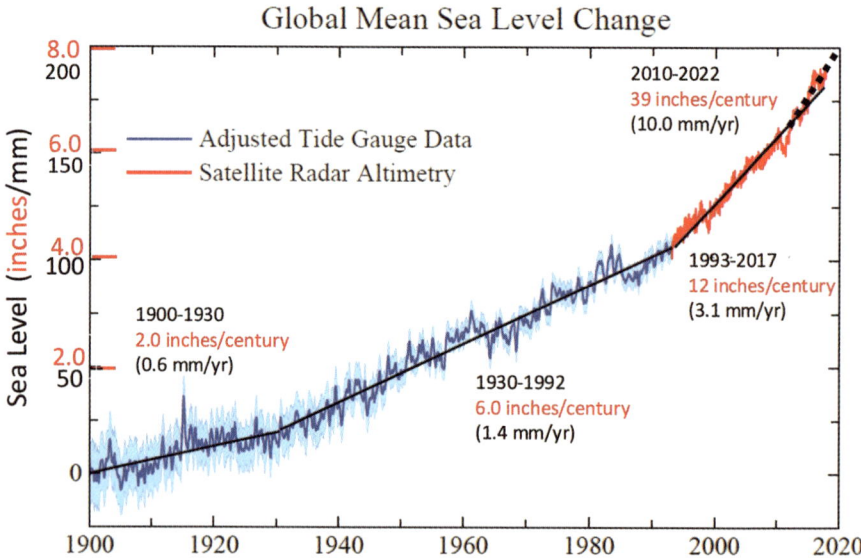

FIGURE 2.5. Global mean sea-level record showing the historical acceleration of sea-level rise. The black dotted line shows the average rate along the US Atlantic and Gulf Coasts since 2010, which has increased to 39 inches/century (10.0 mm/yr) (Dangendorf et al., 2023). Modified from Columbia.edu (James Hansen).

Indeed, this acceleration is known to result from the oceans warming and expanding (*steric influence*) and the accelerated melting of glaciers and ice sheets. The actual rate varies worldwide because of the unequal distribution of heat in the oceans and isostatic changes associated with unloading from retreating ice sheets. Results from a recent study show that since 2010, the average rate of rise in the Gulf of Mexico has surpassed the global average, rising 39 inches/century (~1.0 m/century) (Dangendorf et al., 2023). This matches the rate during the early Holocene when the coast was undergoing rapid change. This more recent acceleration is believed to be driven by changes in ocean circulation patterns, so it is a regional phenomenon and may prove to be of limited duration, although the cause may be related to climate change.

I have lectured on sea-level rise and its impacts on coasts many times, and one question I often get is, "Why should we be concerned about an increase in the rate of rise of only a few inches over decades?" A common assumption is that at this rate, it will be decades before we are faced with the impacts of rising sea levels, so there is plenty of time to address the problem. In reality, the Gulf Coast is responding rapidly to the current acceleration. A significant portion of coastal wetlands will be lost by 2050 (fig. 1.2). This is because the survival of coastal wetlands depends on their capacity for *vertical accretion*, which is the rate at which they can gain elevation by adding new sediment at their surface. Scientists have found that the battle of wetland sustainability versus drowning hinges on just a few inches per century (Morris et al., 2002). South Louisiana is the "poster child" for this scenario because its wetlands are losing this battle. There, the impacts of accelerating sea-level rise are compounded by high subsidence rates and a decrease in sediment supply caused by human alteration of the natural sediment delivery and dispersal system.

Barrier islands and peninsulas are also susceptible to sea-level rise, but their response is complicated and poorly understood. This is because multiple factors, particularly coastal gradients and sand supply, influence shoreline stability during sea-level rise. Numerical models show that low-gradient coasts with low sand supply retreat landward in response to a relatively small increase in the rate of sea-level rise (e.g., Lorenzo-Trueba and Ashton, 2014). Again, the balance between sand supply and sea-level rise along much of the Gulf Coast appears to be tilted in favor of increased shoreline erosion (Anderson et al., 2023).

Causes of Sea-Level Change

The leading causes of global sea-level rise are the warming and thermal expansion of oceans and the melting of glaciers and ice sheets. Another influence is changes in water storage on land, but this effect is considered minor compared to steric and glacial contributions. Changes in the volumes of ocean basins are also an important component, but only over geological timescales.

The oceans are the ultimate sink for a vast percentage (~90 percent) of the excess heat generated by climate warming. As water temperatures increase, oceans expand. This steric influence is predictable because the rate of ocean thermal expansion and sea-level rise relative to surface water temperature increase is well established. Oceanographers have been acquiring ocean temperature profiles for more than a century, and the results are clear: the oceans are getting warmer. Since 1950, surface water temperatures have increased by 0.65°C on average, accounting for roughly one-third of the observed increase in sea-level rise. The amount of sea-level rise resulting from steric influence varies globally because heat distribution in the oceans is quite variable. The Gulf of Mexico gradually shifted to warmer surface water temperatures this past century, with a notable trend toward warmer surface waters during the past three decades (fig. 2.6). Record-breaking

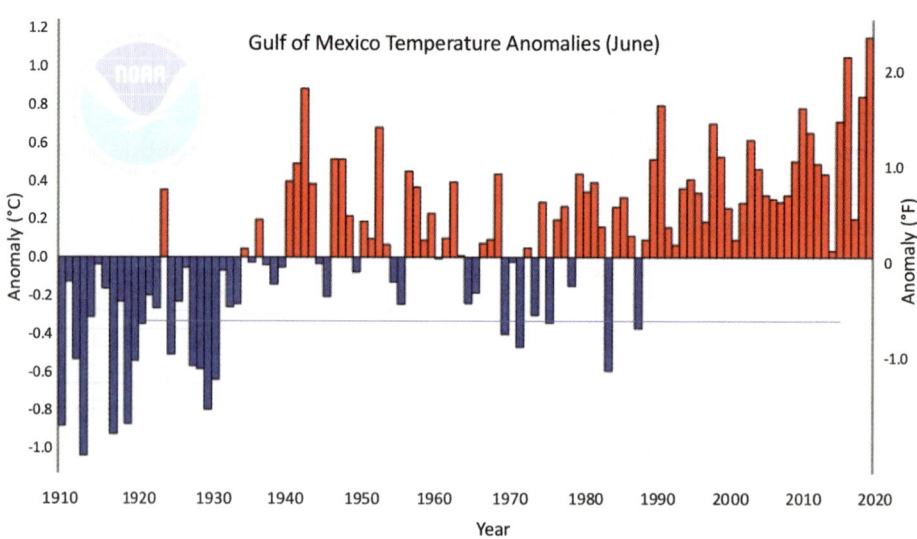

FIGURE 2.6. June sea-surface temperature anomalies in the Gulf of Mexico since 1910. Modified from NOAA.

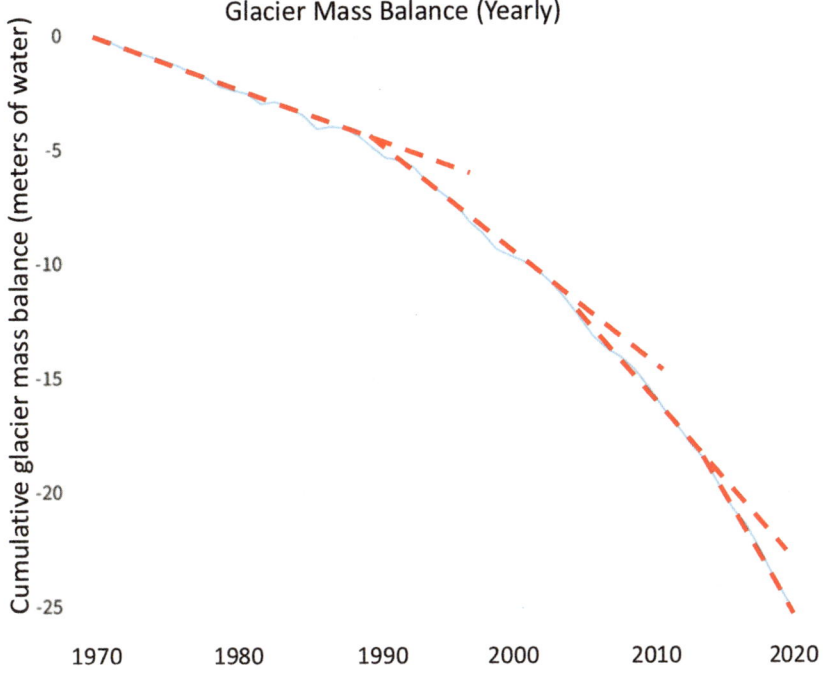

FIGURE 2.7. Ice loss from glaciers since 1970. Modified from the World Glacier Monitoring Service.

temperatures occurred in 2023, peaking at 31°C in August. It is important to know that sea-level response to increasing ocean temperatures occurs rapidly and could continue for decades after temperatures stabilize. This is because of slow mixing within the water column.

The other major cause of accelerated sea-level rise is the increased melting of glaciers and ice sheets. The majority of glaciers are currently retreating or showing signs of accelerated melting in historical time (fig. 2.7). This decline in global ice volume began in the nineteenth century at the end of the *Little Ice Age*, before significant warming as a result of increased greenhouse gas emissions. However, a recent study of 98 percent of the world's glaciers showed that the rate has increased this century, with many retreating at alarming rates (Millan et al., 2022). The actual volume of water released to the oceans by melting glaciers is not easily determined, especially for glaciers terminating in the sea. The *Sixth Assessment Report* of the Intergovernmental Panel on Climate Change (IPCC) estimates that combined steric and glacial contribution to global sea-level rise is roughly two-thirds the total rise. The remaining component of the rise is attributed mainly to ice sheets. The Greenland and Antarctic Ice Sheets are the most significant potential contributors to sea-level rise, but also the most difficult to measure. As recently as 2020, the European Space Agency issued a report indicating that the Greenland and Antarctic Ice Sheets are losing ice six times faster than in the 1990s, consistent with the IPCC's worst-case scenario for future sea-level rise.

Another minor contributor to the historical acceleration of sea-level rise is the amount of water stored in groundwater, lakes, reservoirs, and snowpack. Climate-related changes in water and snow storage on land have been negligible in recent decades. Still, anthropogenic influences have accounted for several tenths of an inch this past century (IPCC *Sixth Assessment Report*). These influences include water impoundment by dams and reservoirs, irrigation, and urban and industrial groundwater usage.

Ice Sheet Contributions

Ice sheets are the most significant potential contributors to sea-level rise. The Antarctic Ice Sheet is large enough to raise sea level by approximately 200 feet, and the Greenland Ice Sheet has enough ice to raise sea level by approximately 25 feet. Thus, only a minor decrease in the volumes of these ice sheets could significantly add to the rise in global sea level. Unfortunately, assessments of the future contribution of ice sheets to rising sea levels remain uncertain.

I began working in Antarctica half a century ago. During my early career, it was generally believed that the Antarctic Ice Sheet did not significantly contribute to rising sea levels. The state of the Greenland Ice Sheet at that time was also poorly understood. It is now widely believed that Antarctica and Greenland are the leading contributors to the rise in sea levels now and in the future. This century may include episodes of retreat when the rate of rise is easily double or triple the current rate. These pulses in the rate of rise will not be distributed uniformly across the globe. The Gulf Coast will experience rates 15–30 percent higher than the global average from West Antarctica (Mitrovica et al., 2009). Dr. Eric Rignot, one of the foremost authorities on ice sheet contributions to sea-level rise, came to the following conclusion in his review of the topic. "In several sectors of Greenland and Antarctica, I conclude that multi-meter sea level rise is inevitable, but the rate of sea level rise will depend on how urgently we keep climate warming under control and subsequently bring the climate system back toward pre-industrial levels" (Rignot, 2021). To better understand the reasons for this concern and why the alarm did not sound sooner, we need to understand the complex behavior of ice sheets and their response to climate change.

The Antarctic Ice Sheet

The Antarctic Ice Sheet occupies a continent centered over the pole, resulting in icy (polar) conditions where surface melting is generally confined to the Antarctic Peninsula and occurs only during the austral summer. Elsewhere, the ice sheet is in a polar state, meaning that melting is limited mainly to the ice sheet's base. This basal melting results from the tremendous pressure caused by the weight of the ice mass, a process known as *pressure melting*, augmented by frictional heat generated as the ice flows across the bed. Antarctica has no rivers, although there is strong evidence for significant subglacial water drainage and many lakes where water is stored. The main form of ice sheet decay is the calving of icebergs that drift away from the continent before melting. Both subglacial water discharge and iceberg calving are challenging to quantify, so Antarctica's contribution to sea-level rise remains elusive.

Antarctica is two continents occupied by two different ice sheets. East Antarctica is a very old continent with relatively high elevation. It is covered by the East Antarctic Ice Sheet (EAIS), which averages nearly two miles thick and is mainly grounded above sea level (fig. 2.8b, profile A-A'). Over large areas of East Antarctica, ice flow lines diverge toward the coast, resulting in deceleration of the ice and relatively slow flow velocities, typically less than 60 feet/year near the coast. There are, however, exceptions to this drainage pattern, such as in the Amery drainage basin and the Wilkes Land–eastern Ross Sea drainage basins, where convergent flow results in faster flow velocities toward the coast. Because of its relatively slow drainage, the EAIS maintains a rather steep profile at its margin.

West Antarctica is an archipelago with large islands and continental fragments covered by the West Antarctic Ice Sheet (WAIS). The WAIS is mainly grounded below sea level, up to a mile below sea level near its center (fig. 2.8c, profile B-B'). As the ice sheet flows seaward into deep marginal basins, its flow lines converge, and flow speeds accelerate to form rapidly flowing ice streams. These ice streams flow at velocities of hundreds to thousands of feet per year and account for the majority of ice discharge from the continent. This rapid ice drainage results in a thinner ice sheet

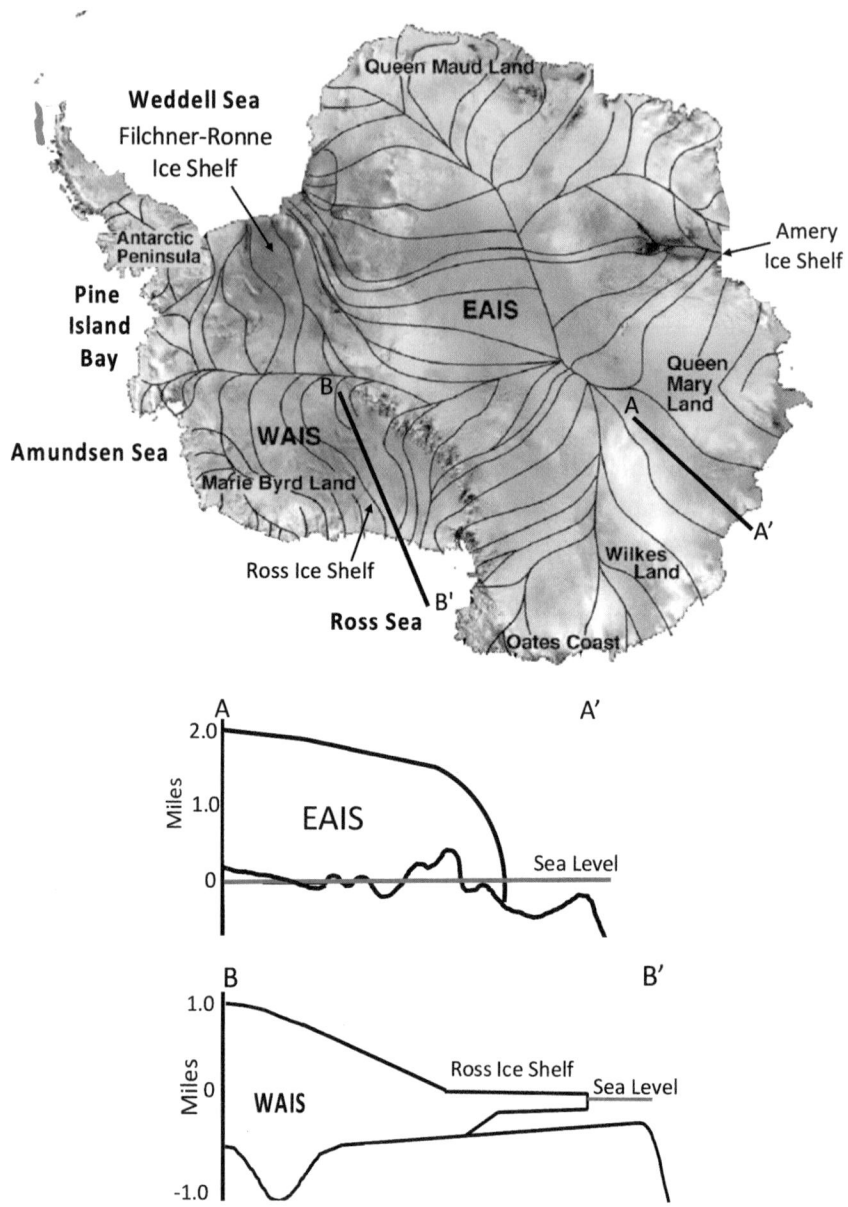

FIGURE 2.8. (a) Drainage map of the Antarctic Ice Sheet, which is divided into the East Antarctic Ice Sheet (EAIS) and West Antarctic Ice Sheet (WAIS). Lines show ice flow directions. Note that the drainage from the EAIS is mostly divergent toward the coast, which decreases flow velocities. Exceptions are drainage into the Amery Ice Shelf and Wilkes Land coast. West Antarctica is dominated by the convergence of ice flowing from the ice sheet's interior to the coast, resulting in faster flow and discharge. These differences in ice flow velocity result in (b) a relatively steep marginal profile for the EAIS (profile A-A'), versus (c) the low, thin marginal profile of the WAIS (profile B-B'). Modified from Anderson et al. (2002).

with a lower profile relative to the EAIS. Because of their rapid flow and discharge, ice streams can contribute to sea-level rise faster than other portions of the ice sheet.

Portions of the EAIS and virtually all of the WAIS are grounded below sea level at their margins. The boundary between grounded and floating ice is called the grounding line. It is where ice is thick enough to remain grounded if relative sea level and ice thickness do not change significantly. Seaward of the grounding line, ice shelves exist where the ice is thin enough to float. The most extensive ice shelves are the Ross Ice Shelf and the Filchner-Ronne Ice Shelf (fig. 2.8a). Because the WAIS has a thinner ice margin, it has a more extensive marginal zone where the ice thickness is close to buoyancy. As a result, it is considered relatively unstable, especially given its rapid flow and ice discharge rate. If the margin of the ice sheet becomes thinner, or if sea level rises, the ice sheet will float off its bed and retreat landward. An increase in flow velocity can lead to a decrease in ice thickness near the grounding line, ultimately leading to grounding line retreat. Once detached from the bed, the ice can flow faster, and ice discharge from the continent increases. That discharge occurs mainly by fracturing and breakup of the ice shelf to produce icebergs. In contrast, the relatively steep profile around much of the EAIS, coupled with its generally slower flow and discharge, results in a more stable ice sheet, meaning the ice sheet grounding line is less sensitive to marginal thinning or sea-level rise.

Over four decades ago, Dr. Terry Hughes of the University of Maine advocated the idea that Pine Island Bay in the Amundsen Sea (fig. 2.8a) was the "weak underbelly" of the West Antarctic Ice Sheet. His logic centered on the fact that the two largest glaciers draining into Pine Island Bay, Thwaites and Pine Island Glaciers, together accounted for about one-third of the ice discharge from West Antarctica. He also believed that their flow and discharge were accelerating. While the logic behind Dr. Hughes's concern was well grounded, it was virtually impossible at that time to test his hypothesis. Since then, the concept of marine-terminating glaciers and collapsing ice streams has been widely accepted by glaciologists. This change was driven mainly by new technology for investigating glaciers and ice sheets from space, on the ice, and from the sea. Improvements in satellite measuring methods have been essential, particularly NASA's ICESat, which uses laser and radar altimetry obtained by the ESA-s, ERS-2, and CryoSat-2 satellites. Data from these satellites allow detection of small variations in the surface of ice sheets across individual ice drainage basins over years to decades, depending on the rate of change. These new data have revealed rapid changes in Thwaites Glacier, making it the largest potential contributor to future sea-level rise.

Research in recent decades has led to the discovery that the rapid retreat of Thwaites Glacier is driven by the breakup of its floating margin and that relatively warm ocean waters (~0°C) flowing into Pine Island Bay are providing heat that is melting the cold (~ −8.0°C) base of the ice shelf. As a result, ice loss from Thwaites Glacier has increased fivefold in the last three decades. Currently, this accounts for about 4 percent of global sea-level rise. But, depending on how fast and how much ice is ultimately discharged into the sea from Thwaites Glacier, it could contribute as much as two feet of sea-level rise within a century. Given current rates of ice volume loss, the grounding line in some locations could begin to float within a matter of decades, resulting in even faster sea-level rise. The rate of ice discharge depends on several factors that are currently the focus of considerable research.

Given its vast size and thickness, the Antarctic Ice Sheet depresses the continent on which it rests, a process known as *glacial isostasy*. Because the thickness of the ice decreases away from its center, the result is a bowl-like isostatic depression beneath the ice sheet that extends to the continental shelf (fig. 2.9). The landward slope of the bed has been further accentuated by repeated glacial advances that have eroded inland areas and deposited a thick pile of sediments on the outer continental shelf. As a result, the sea-floor on which the ice sheet rests slopes landward. In 1978, two prominent glaciologists, Dr. Robert Thomas and Dr. Charles Bentley, published a land-mark paper in which they argued that this configuration of a lens-shaped ice sheet resting on a landward-sloping bed contributes to the instability of the ice sheet and could result in a runaway condition where grounding line retreat accelerates as the ice sheet retreats into deeper water (Thomas and Bentley, 1978). That same year, Dr. John Mercer of Ohio State University published a similar ice sheet collapse model suggesting that climate change resulting from rising greenhouse gases might trigger ice sheet thinning and collapse. Mercer noted, "If the greenhouse warming effect of the resultant increasing atmospheric CO_2 *is as great as the most advanced current models suggest*, a critical level of warmth will have been passed in high southern latitudes 50 years from now, and deglaciation of West Antarctica will be imminent or in progress. Deglaciation would probably be rapid once it had

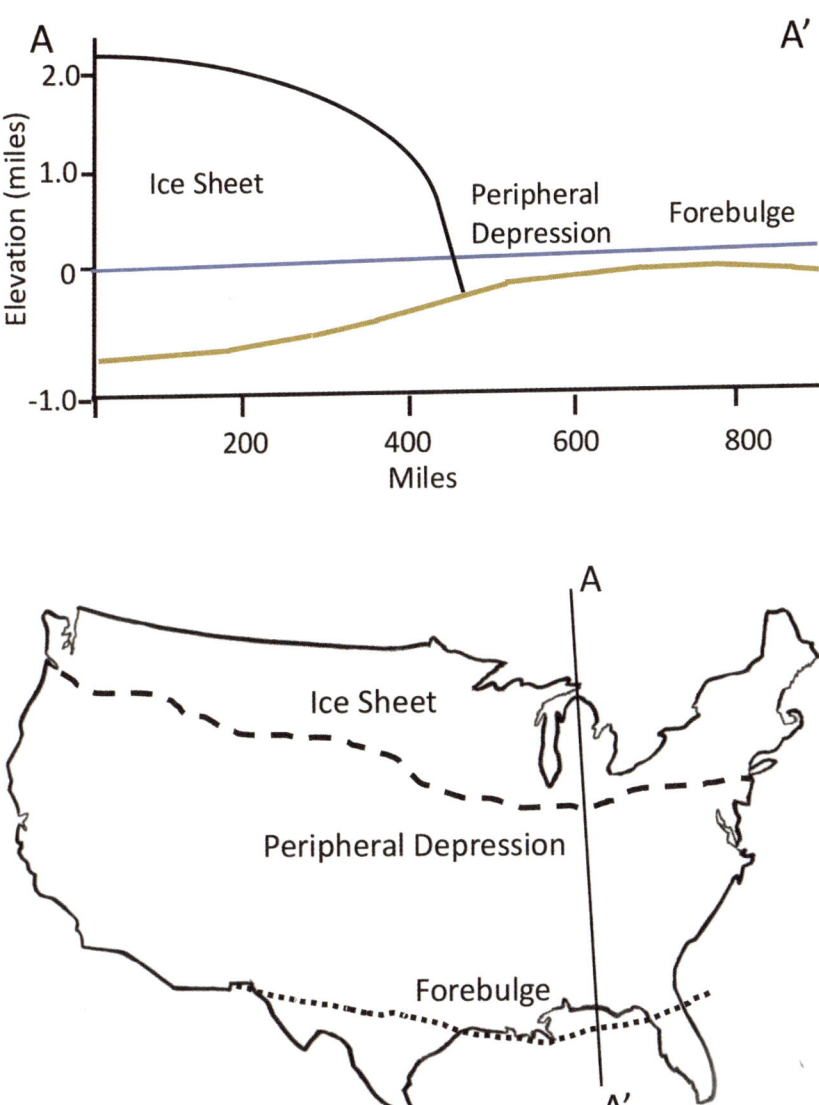

FIGURE 2.9. Isostasy associated with ice sheets. The land beneath the ice sheet is depressed under the weight of the ice by roughly one-third the thickness of the ice, while a peripheral depression extends about 200 miles from the ice margin. Seaward of the peripheral depression, the land or seafloor surface is forced upward as an isostatic forebulge. When the ice sheet retreats, the isostatic and peripheral depression rebound while the forebulge relaxes and sinks. The map shows the estimated extent of the North American forebulge during the LGM.

started, and when complete, it would have led to a rise in sea level of about 5m along most coasts" (Mercer, 1978).

By the late 1980s, new satellite data and fieldwork provided compelling evidence that ice streams flow much faster than previously known and play a vital role in the ablation of the Antarctic and Greenland Ice Sheets. This is a familiar story in science; observations often remain unexplained until new technology opens the door for discovery. In this case, observations from polar-orbiting satellites led to this discovery. One of the most intriguing questions concerns the mechanism of rapid ice stream flow.

Ice sheets generally flow at velocities of tens of feet per year, which is the case for much of the EAIS. So, how do we explain much faster rates of ice stream flow? To attain high velocities, the ice must be sliding across the bed on which it rests. However, it was argued that in the case of ice streams, this would require widespread water dispersal at the bed, a condition not supported by either theory or observation. It would be another decade after Thomas and Bentley's ideas were published before two of Dr. Bentley's PhD students at the University of Wisconsin, Richard Alley and Don Blankenship, published their seminal paper suggesting that the rapid flow of ice streams is a result of ice sliding across a slippery mixture of water and sediment, known as a *deforming bed* (Alley et al., 1986). This theory has since been tested by drilling through an ice stream and sampling the deforming bed.

It is now widely accepted that the behavior of ice streams, which are widely recognized as having a significant influence on ice sheet stability, is strongly influenced by the nature of the bed on which the ice sheet rests. In general, sliding velocities are relatively slow where the ice sheet flows across hard bedrock, and relatively fast where it flows over sedimentary beds. These sedimentary beds are eroded and mixed with water to form the 'deforming bed on which ice streams slide. Subsequent discoveries of hundreds of large lakes beneath the ice sheet provided evidence for the large quantities of water needed to maintain a widespread deforming bed. At first, it was unknown how this water could be widely dispersed. Then, seafloor images from the continental shelf revealed elaborate channel networks connecting subglacial basins. These combined discoveries offer compelling evidence for subglacial plumbing systems capable of dispersing subglacial water to the bed in quantities large enough to sustain ice streams. But they also revealed a complex process that could increase or decrease in response to multiple factors, thus influencing ice stream behavior.

In recent decades, more research has focused on whether large ice streams in Antarctica and the Northern Hemisphere experienced rapid change. If so, rapid ice sheet retreat could be the mechanism that caused the punctuated sea-level rise events of the early Holocene, and it could even contribute to future sea-level rise. Again, new technology paved the way to address this question. In this case, developments in swath bathymetry since the 1990s led to the acquisition of spectacular images of the Arctic and Antarctic seafloor where ice sheets were grounded during the LGM (fig. 2.10a). These images provided our first view of the ice/bed interface, commonly referred to as the death mask of the ice sheet. Some of the most compelling images show *recessional moraines* formed during the rapid retreat of the grounding line, and wedges of sediment (*grounding zone wedges*) that formed when the grounding line stabilized for decades to centuries (fig. 2.10b). Research in the Ross Sea has resulted in detailed reconstructions of rapid grounding line retreats that likely contributed to rapid sea-level

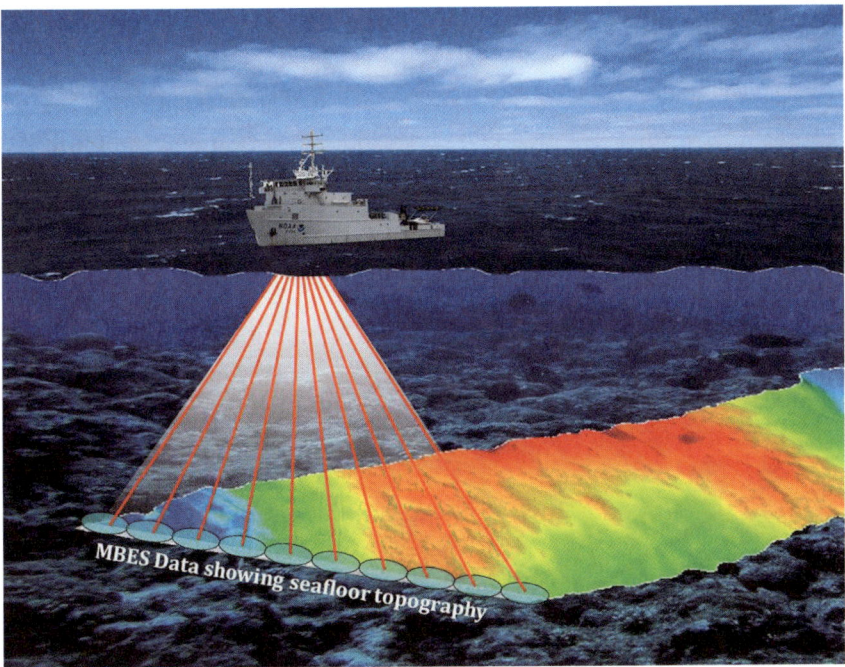

MBES Data showing seafloor topography

FIGURE 2.10. (a) Illustration showing acquisition of swath bathymetry data used to image the seafloor. Image from NOAA.

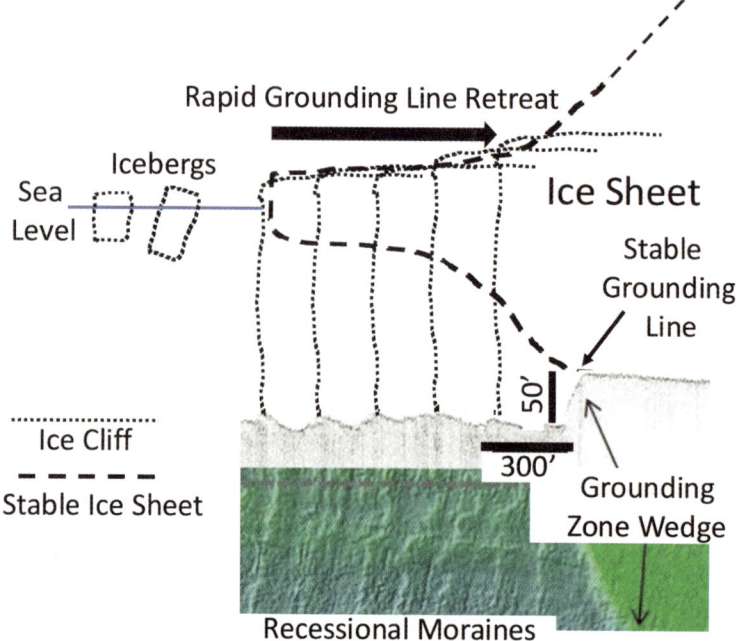

FIGURE 2.10. (b) Swath bathymetry image shows recessional moraines, indicative of stepwise grounding line retreat that culminated in a period of grounding line stability and deposition of a large grounding zone wedge. Rapid grounding line retreat results in ice shelf collapse and the formation of a vertical ice margin (ice cliff). Grounding line stability results in ice shelf formation and deposition of a wedge of sediments beneath the grounding zone.

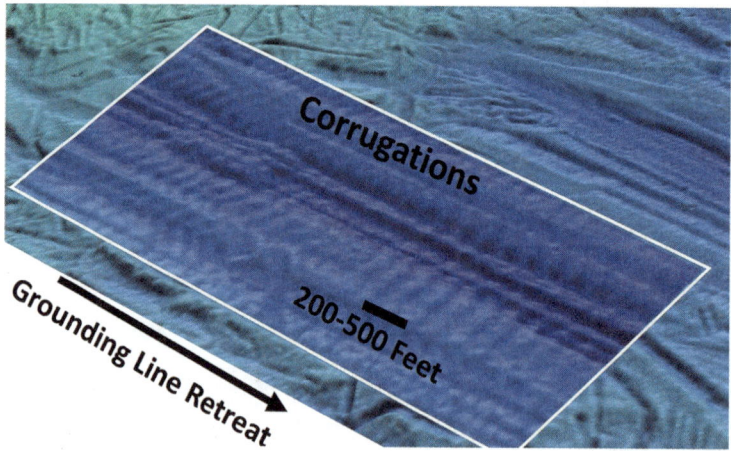

FIGURE 2.10. (c) High-resolution seafloor image showing small-scale grounding line features (corrugations) that average 3 to 5 feet in height, which matches the tidal amplitude. Ridge spacing suggests rapid grounding line retreat rates of 60 to 80 feet daily.

rise (Bart and Kratochvil, 2022). Other studies have revealed fairly equally spaced ridges, called *corrugations*, with three-to-five-foot amplitudes, consistent with grounding line retreat regulated by ocean tides (Jakobsson et al., 2011; Graham et al., 2022). Thus, corrugations indicate the grounding line's sensitivity to sea-level fluctuations of a few feet. They are commonly overprinted by deep *iceberg furrows* formed during the final breakup of an ice shelf and rapid grounding line retreat (Yokoyama et al., 2016).

The Greenland Ice Sheet

The Greenland Ice Sheet (GIS) is the sole surviving ice sheet in the Northern Hemisphere. It is nearly two miles thick near its center and contains enough ice to raise global sea level by approximately 25 feet if completely melted. Of more immediate concern is that the ice sheet could contribute as much as 1.5 feet of sea-level rise by the end of this century.

Unlike the Antarctic Ice Sheet, the GIS has retreated inland from the continental shelf. Also, significantly less ice is discharged to the sea through ice streams. The exceptions are a few large glaciers that flow into fjords, but there is little direct marine influence on the ice sheet. However, the ice sheet has a recent history of increased melting, and there are strong indications that its flow is now accelerating in response to increased atmospheric warming. Ice sheet melting during the past two decades was six times faster than in the 1980s, triggered by atmospheric temperatures that are the warmest in a thousand years (Hörhold et al., 2023). Widespread surface melting is marked by expansive areas of icy slush, meltwater ponds, and rivers of meltwater that disappear where the water flows through cavities down to the bed, promoting basal sliding and increased discharge to the sea (fig. 2.11). Thus, unlike the Antarctic Ice Sheet, where melting occurs on the underside of the ice sheet, the surface melting and retreat of the GIS is more conspicuous. The ice sheet's retreat can be measured using satellite images that show the ice margin and new bedrock exposures (fig. 2.12). Furthermore, there is growing consensus among scientists that temperatures in Greenland could increase by 3°C by the end of this century. Based on ice sheet models, that would result in significant mass wasting, if not the complete demise of the ice sheet.

In summary, our understanding of modern ice sheets has grown exponentially in recent decades, mainly thanks to new technology for observing the ice sheets from space and imaging the former bed of the ice sheets where they were grounded on the continental shelf. These seafloor images

FIGURE 2.11. Photograph of the Greenland Ice Sheet showing meltwater streams flowing across the surface to the sea. Image from NASA.

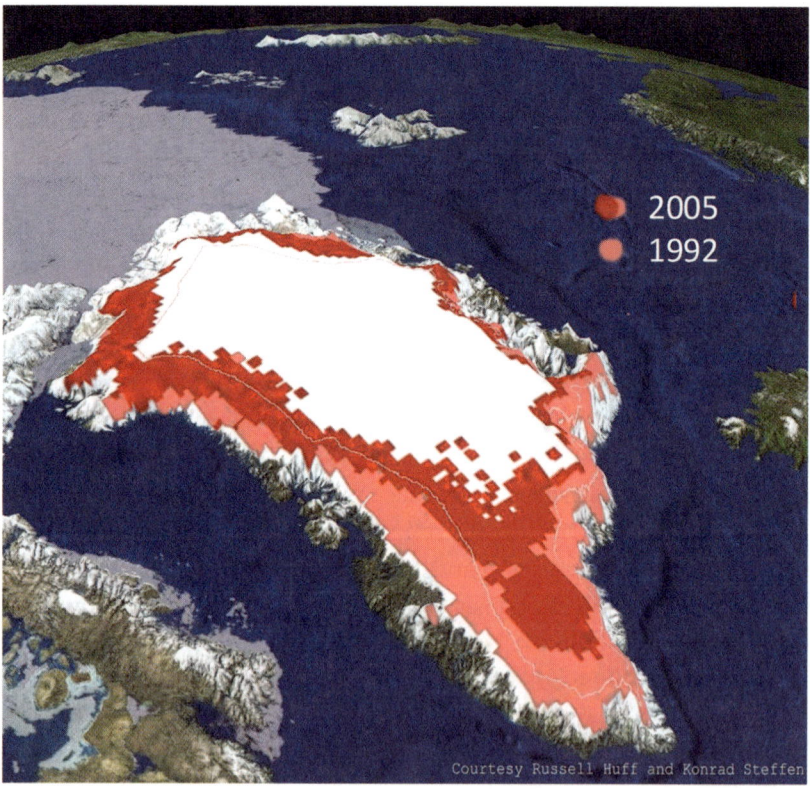

FIGURE 2.12. Satellite image of the Greenland Ice Sheet showing areas of significant surface melting between 1992 and 2005. From CIRES 1, University of Colorado.

have yielded compelling evidence for periods of rapid ice sheet retreat that likely resulted in punctuated sea-level rise (fig. 2.10). There is emerging evidence that the ice sheets are again experiencing instability that could lead to similar punctuated sea-level rise in the future. For example, Thwaites Glacier has become known as *Doomsday Glacier* because it is believed to be on the verge of collapse, which could result in a sea level rise of several inches within decades.

Subsidence

The term "subsidence" refers to the lowering of the land surface relative to some datum. Subsidence is caused by a number of processes, including *tectonic subsidence* of the Gulf of Mexico Basin, *glacial isostatic adjustment* (GIA) resulting from the retreat of the North American Ice Sheet and collapse of its isostatic forebulge (fig. 2.9), sediment compaction, faulting, and subsurface salt migration (fig. 2.13). Scientists have made significant headway in understanding these processes, but major challenges remain in measuring subsidence rates and determining the relative influence of the different processes involved. This is critical research because the degree to which a particular portion of the coast will be impacted by sea-level rise depends on establishing the subsidence rate and the sediment supply needed to compensate for this rise.

Basin subsidence is a natural response to the tectonic lowering of the Gulf of Mexico Basin and loading by the thick column of sediments that fill the basin (fig. 2.13). The western Gulf Coast has a long history of relatively high subsidence driven by the greater thickness of sediments within the basin relative to the more stable Florida Platform. Research indicates that the combined tectonic subsidence and subsidence from sediment loading, or *sediment isostatic adjustment* (SIA), is less than 8.0 inches/century (2.0 mm/yr) along the Gulf Coast. The highest rates are associated with deltas in Louisiana and Texas.

During the LGM, the Laurentide Ice Sheet extended across much of the northeastern United States and attained a total thickness of 1.5 to 1.9 miles. The weight of the ice sheet resulted in an isostatic depression of several hundred feet that extended approximately 125 miles beyond the ice sheet margin (fig. 2.9). South of this depression, the isostatic forebulge extended several hundred miles into the north-central Gulf Coast. After the LGM, the ice sheet retreated to the north, and the release of its great

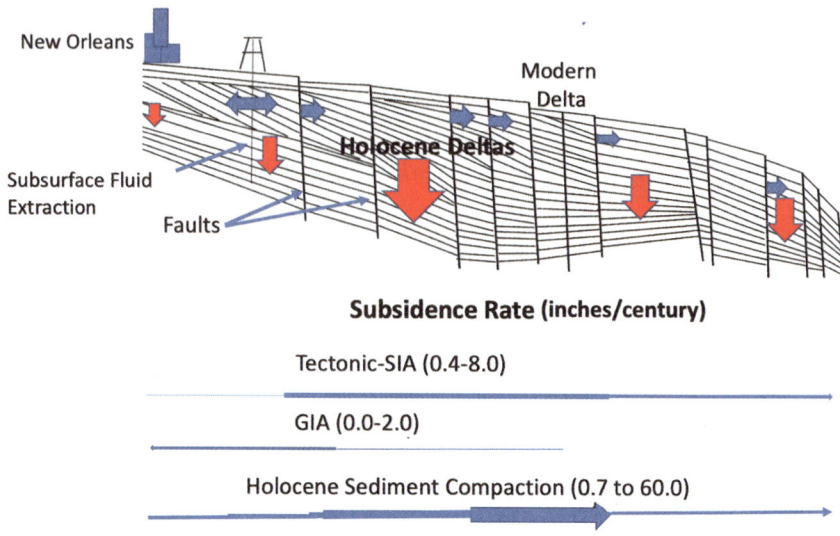

FIGURE 2.13. An idealized stratigraphic profile extending from New Orleans to the edge of the Louisiana continental shelf illustrating different subsidence and rate modes. The overall trend is one of thicker sediments in an offshore direction, which results from tectonic subsidence of the Gulf of Mexico Basin. Tectonic subsidence and subsidence from sediment loading (SIA) are greatest near the modern coast, where thicknesses are greatest. Glacial isostatic adjustments (GIA) caused by forebulge collapse account for between 0.0 and 2.0 inches/century of the total subsidence in south Louisiana and are negligible elsewhere along the coast. The most significant component of subsidence is the compaction of Holocene sediments, which can be as high as 60 inches/century. Other compaction modes include faulting and fluid extraction, which are more localized. Bold red arrows designate relative rates of subsidence.

weight resulted in isostatic rebound of the land surface that continues today in portions of Canada and the northeastern United States. Meanwhile, forebulge collapse occurs in south Louisiana where GIA is estimated to be between 0 and 2.0 inches/century (0 and 0.5 mm/yr). Elsewhere along the Gulf Coast, GIA is believed to be negligible.

Measuring Sea-Level Rise

Historical rates of relative sea-level rise are measured using tide-gauge records, which in Europe extend back to ~1700. In the United States, the oldest record, from Galveston, Texas, extends back to 1900. The gauges are commonly placed on piers and measure sea level relative to a nearby *geodetic benchmark*. The drawback to this approach is that the gauges may

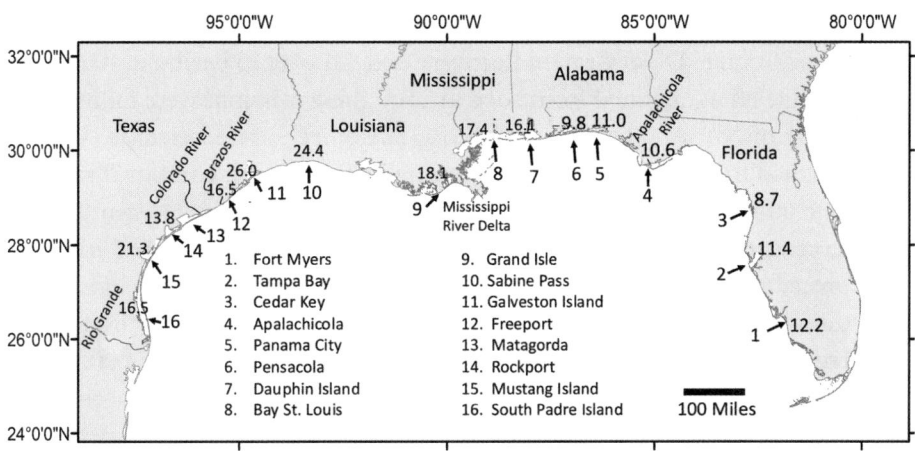

FIGURE 2.14. Relative sea-level rise (RSLR) in inches per century measured by tide gauges across the Gulf Coast. The numbered stations designate locations referred to in the text. Mobile Bay is the dividing line between average rates of more than 16.1 inches/ century to the west and less than 12.2 inches/century to the east. This difference reflects the greater contribution of subsidence to RSLR in the western Gulf, which is associated with the Gulf of Mexico Basin. More localized variations are mostly the result of differences in the thickness of Holocene sediments. Hundreds of data points exist for south Louisiana, indicating coastal subsidence between 12 and 100 inches/century. Modified from Anderson et al. (2023).

move vertically with subsidence or faulting, so they measure relative sea-level rise (RSLR). As a result, there is considerable variability in actual tide-gauge measurements across the Gulf Coast, with Mobile Bay being the dividing line between higher rates to the west and lower rates to the east (fig. 2.14). This difference reflects the more significant contribution of subsidence to RSLR in the western Gulf. Because of this, tide gauges in areas with little or no subsidence, such as coastal Florida, are used to measure the *mean rate of sea-level rise* (MSLR) since observations began in the early 1920s. That rate is ~11 inches/century (28 cm/century) and is close to the global average.

Satellite altimetry provides much improved sea-level records because it measures changes in sea surface elevation and ignores subsidence. This is because satellites use gravity or radar altimetry to measure the distance between the sea surface height and the center of the Earth, known as geocentric sea level. The first satellite measurements of sea level were from the TOPEX and Jason satellite missions, which began in 1992. This was followed by satellite measurements from a pair of NASA–German Aerospace Center (DLR) Gravity Recovery and Climate Experiment (GRACE) satellites that operated from 2002 to 2017. These missions were followed by a pair of NASA–GFZ German Research Centre for Geosciences GRACE Follow-On satellite missions beginning in 2018. These combined satellite data indicate an average global rate of sea-level rise of 12 inches/century (0.3 m/century) since 1993, which is significantly higher than the rate during the preceding century and a half (fig. 2.5). Despite these advances in satellite measurements, there has been insufficient spatial resolution to measure regional variability in sea-level change, especially variations in land subsidence, which can vary significantly over a few miles. That has changed with the launch of NASA's ICESat-2 satellite, which uses light detection and ranging (*LiDAR*) coupled with a scanner and specialized *Global Positioning System* (GPS) receiver to provide measurements accurate enough to allow three-dimensional surface elevation mapping of the ocean and land surface. It provides the level of detail needed to address coastal response to sea-level rise.

Geological and Archaeological Records of Sea-Level Rise

The previous debate about the historical acceleration of sea-level rise was fueled by the lack of tide-gauge records that extended far enough back to yield a compelling record of this acceleration. Satellite data have resolved this conflict, but there are also ways to extend the record further back in time using archaeological records. Some of the best examples come from Venice, Italy, where rising seas and subsidence have resulted in annual flooding events within the city (Zanchettin et al., 2020). Landscape paintings were used to measure changes in algae growth between 1571 and 1758. The results showed that RSLR was 0.5 inch/century (12.7 mm/century) during this time interval. Another study focused on the submergence of palace steps on Venice's central canal dating back to 1350. This study showed that during the mid-fifteenth century, sea level rose 3.9 to 6.8 inches and then decreased to a minimum around 1730. This later episode of slow sea-level rise coincided with the Little Ice Age, a period of global cooling and associated glacial advance. Since the Little Ice Age, the rate of rise has increased.

Other evidence for a historical acceleration in the rate of sea-level rise has come from geological records for the past millennium. Figure 2.15 shows one example from the North Carolina and South Carolina coasts that indicates an increase in the rate of rise that began in the late 1800s

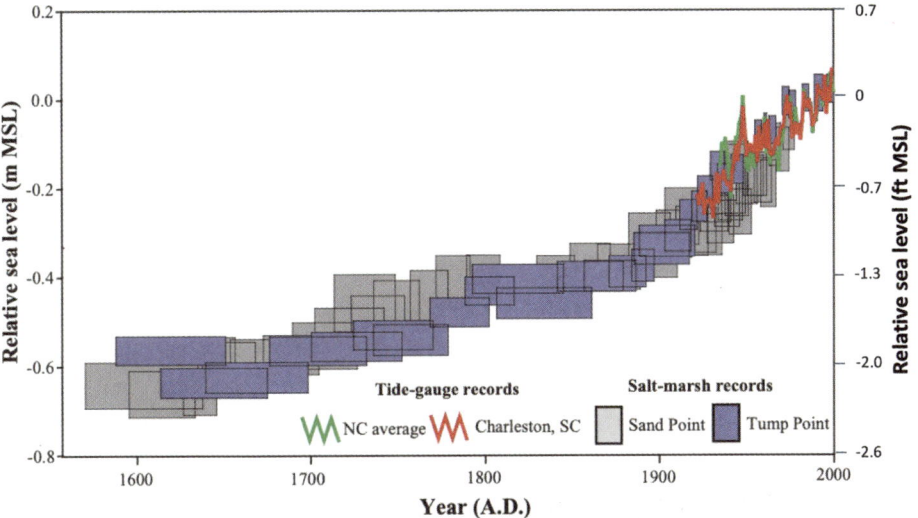

FIGURE 2.15. Sea-level records constructed from coastal marsh deposits of North Carolina and South Carolina. Note that the acceleration in the rate of rise began in the late 1800s (Kemp et al., 2009). Reprinted with permission from the Geological Society of America.

(Kemp et al., 2009). Similar studies from other areas have since supported these results. One example is a study in Florida saltmarshes, which indicate a similar increase in the rate of rise beginning around 1750 (Gerlach et al., 2017). The sea-level curve for the western Gulf of Mexico (fig. 1.4b) shows a long-term trend with an average rate of rise over the past ~4,000 years of 2.0 inches/century, compared to the current rate of ~11 inches/century derived from tide-gauge records and satellite data.

Measuring Subsidence

Before the early 2000s, subsidence studies in the Gulf Coast region relied on tide-gauge records, with the rate being determined by subtracting the measured rate from the global average rate of sea-level rise. South Louisiana's results yielded significantly faster rates than the global rate. This led to research aimed at using other methods to measure RSLR in the region. One of the first comprehensive studies was conducted by Dr. Kurt Shinkle and Dr. Roy Dokka, who measured elevation changes using hundreds of geodetic benchmarks across Louisiana and neighboring states. These benchmarks were attached to rods anchored at various depths below the ground surface, allowing for subsidence measurements at depth within the sediment column. Their lower Mississippi River Delta measurements ranged from 20 to 100 inches/century. The publication of these results in 2004 led to debate about the reliability of these measurements and the causes of subsidence. This debate was fueled by arguments that oil and gas extraction contributed to high RSLR rates. During the next two decades, significant progress was made in addressing these discrepancies using more precise measuring technologies and additional scientific analysis.

A 2019 report issued by Louisiana's Coastal Protection and Restoration Authority presented results from GPS measurements in the Terrebonne Basin showing subsidence rates between 0.7 and 2.3 feet/century (fig. 2.16) (Byrnes et al., 2019). Outliers remain where subsidence rates are significantly higher, but these are believed to be associated with near-surface faulting and subsurface fluid extraction. Dr. Torbjörn Törnqvist and his

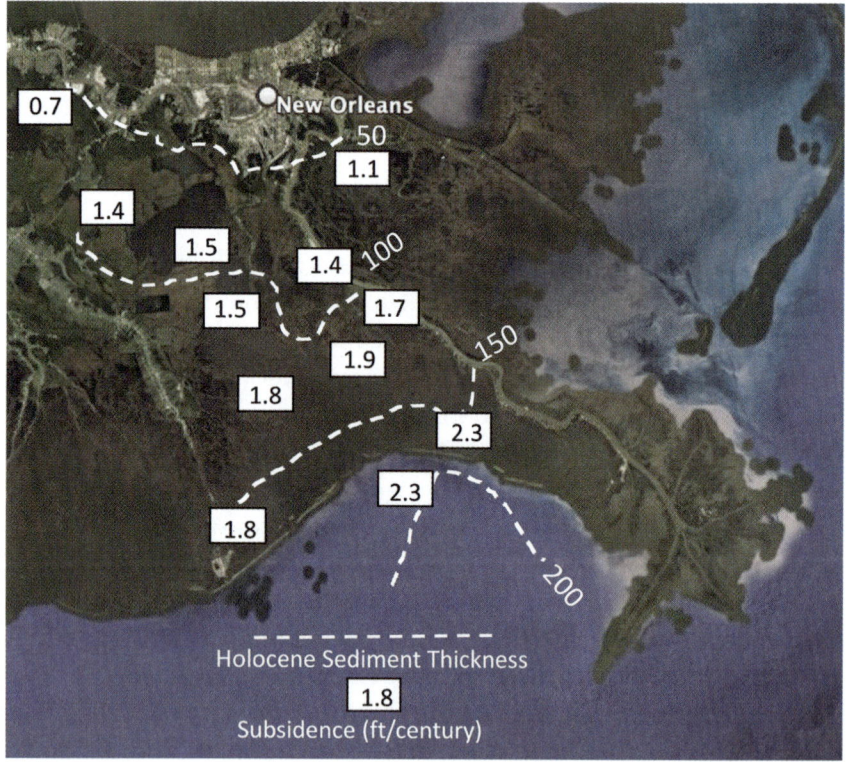

FIGURE 2.16. Subsidence measurements and thickness of Holocene sediments for the Terrebonne Basin in south Louisiana. Note that rates increase toward the south as Holocene sediment thickness increases. Subsidence rates from Byrnes et al. (2019). Base map from Google Earth.

students and postdocs at Tulane University have provided convincing evidence that subsidence results mainly from compaction of unconsolidated, water-saturated Holocene sediments (Törnqvist et al., 2008). There is a general pattern in which subsidence rates increase from north to south as the thickness of Holocene sediments increases (fig. 2.16). Similar trends have been observed in other large deltaic settings, including the Yellow River delta and Mekong delta.

The importance of the compaction of Holocene sediments to overall RSLR is observed in other areas of the Gulf Coast. The Tulane group has measured variable subsidence rates on the Louisiana Chenier Plain, located west of the Mississippi Delta, in the range of those measured in the Terrebonne Basin. In Texas, subsidence rates measured from tide-gauge records vary considerably along the coast, with the highest rates recorded within estuaries known to contain thick Holocene sediments (Anderson et al., 2022). Much lower rates have been recorded in adjacent coastal areas where Holocene sediments are thin to absent.

One of the most contentious debates about RSLR centers around the argument that subsidence in south Louisiana is exacerbated by human activity, specifically oil and gas extraction. The argument against high subsidence rates being the result of fluid withdrawal was that the oil and gas reservoirs were too thin and too deep in the subsurface to cause significant subsidence. This argument was met with the publication of a graph showing that the greatest oil and gas production between 1960 and 1980 in south Louisiana correlated closely to increased land loss in the region (fig. 2.17). This graph also showed that the period of rapid subsidence was followed by a decrease in subsidence to background levels within one to two decades. Notably, the subsidence associated with fluid withdrawal was restricted to areas with significant oil and gas production. It did not account for the increased land loss in other parts of south Louisiana.

The greater New Orleans metro area has also experienced rapid subsidence tied to groundwater and surface water usage and drainage alteration. The NASA Jet Propulsion Laboratory has been monitoring subsidence in the area using airborne radar. That work has revealed significant subsidence in the upper and lower Ninth Ward and Metairie areas and along levees near the Bonnet Carré Spillway.

Land loss resulting from subsurface fluid extraction is also documented in other areas of the Gulf Coast. The best example occurred along the western shore of Galveston Bay, Texas, where discrete episodes of rapid subsid-

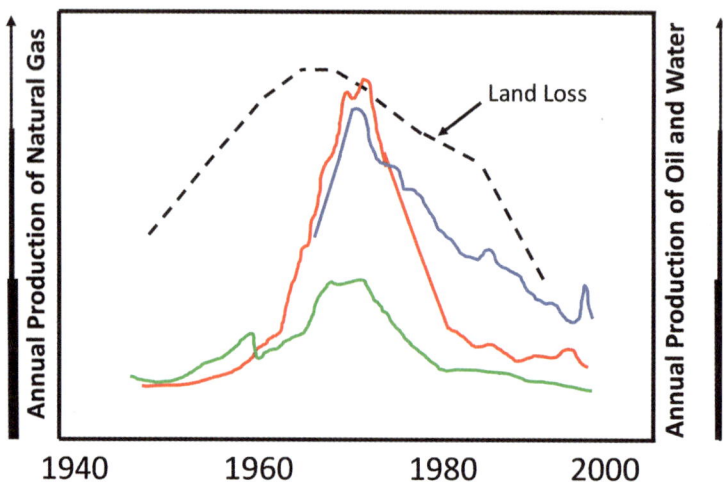

FIGURE 2.17. Natural gas (red line), oil (green line), and water (blue line) withdrawal relative to coastal land loss (dashed line) in Louisiana. From Morton et al. (2002). The most significant land loss is correlated with the greatest oil and gas production between 1960 and 1980. This period of rapid subsidence was followed by a decrease in subsidence to background levels within one to two decades. Figure reproduced with permission of the Gulf Coast Association of Geological Societies.

ence have been tied to oil and gas and groundwater extraction dating back to the early 1900s. The most extreme episode occurred along the Houston Ship Channel during the 1940s to mid-1970s, when increased groundwater usage for industrial purposes resulted in up to 7.5 feet of subsidence in the heavily industrialized area surrounding the Houston Ship Channel, including the Clear Lake area where NASA's Johnson Space Flight Center is located (fig. 2.18).

Decadal Sea-Level Oscillations

Tide gauges measure seasonal, annual, and decadal sea-level variations resulting from astronomical, meteorological, and oceanographic tides (fig. 2.19). These higher-frequency fluctuations overprint the long-term rate of sea-level rise, but there are ways to filter these data to isolate the different components of sea-level rise and fall. *Astronomical tides* are predictable, so they can be subtracted from these records. *Meteorological tides* can vary with changing wind patterns. They can be analyzed using historical meteorological data, which are readily available, to distinguish tidal anomalies associated with periods of intense and persistent winds.

FIGURE 2.18. *Facing top*, Subsidence map for the region from Greater Houston to Galveston, Texas, showing subsidence areas that occurred mainly during the 1940s to mid-1970s that were caused by excessive groundwater usage for industrial purposes. Up to 7.5 feet of subsidence occurred in the area of the Houston Industrial Complex on upper Galveston Bay. Contours in feet. From Gabrysch and Bonnet (1975). Credit: US Geological Survey.

Oceanographic tides are the least understood; they result from changes in ocean circulation and are largely unpredictable because time-series oceanographic data are limited. One approach is to "tune" tide-gauge records by removing the more predictable astronomical and meteorological components from the record and assuming that the remaining high-frequency fluctuations record oceanographic tides resulting from changes in ocean circulation.

One of the known oceanographic sea-level influences results from variations in the Gulf *Loop Current*, which can force water onshore or offshore, resulting in sea-level oscillations. The Loop Current is driven as waters

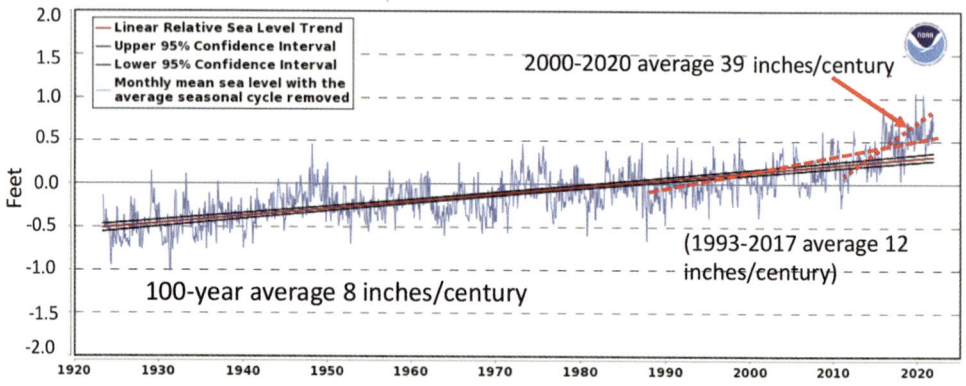

FIGURE 2.19. *Facing bottom,* The tide-gauge record for Pensacola, Florida, shows seasonal, annual, and decadal sea-level variations resulting from astronomical, meteorological, and oceanographic tides. This record is also an excellent example of the slow rise in sea level over the past century, from an average rate of 8 inches/century to 12 inches/century during the satellite measuring era (1993–2017), which is close to the global average. The rate in the Gulf of Mexico has increased to 39 inches/century since 2020, likely because of oceanographic influence. Modified from NOAA.

entering the Gulf of Mexico through the Yucatán Channel flow across the Gulf and exit through the Florida Strait, eventually forming the Gulf Stream (fig. 2.20). The northern extent of the Loop Current varies from year to year and portions of it often break away, forming eddies that can migrate onshore and cause more localized sea-level fluctuations. In Texas, these events are believed to have caused as much as four inches of sea-level rise, potentially influencing monthly to seasonal changes in shoreline migration.

It remains to be seen whether and how climate change influences oceanographic circulation patterns and sea levels in the Gulf of Mexico. The latest-generation satellites are expected to yield a more precise record of how the Loop Current responds to meteorological effects and surface water heating, observations that are needed to test numerical models.

Predicting Sea-Level Rise

This century, the average rate of sea-level rise at Pensacola, Florida, has been 15 inches/century and is close to the global average (fig. 2.19). The most recent analysis of data for the United States, published in 2022 in

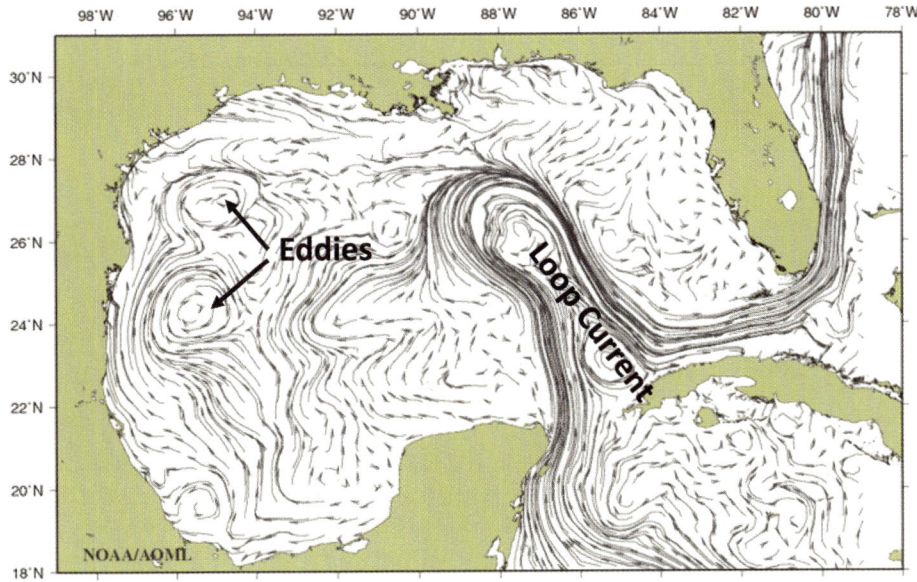

FIGURE 2.20. Generalized wind-driven circulation patterns for the Gulf of Mexico derived from the Intra-Americas Sea Ocean Nowcast/Forecast System (IASNFS) dataset. Modified from NOAA.

NOAA's *Sea Level Rise Technical Report*, revealed that the rate has increased to as much as 39 inches/century during the past two decades. This recent acceleration is likely the result of oceanographic influences.

Results from south Louisiana, where the gap between eustatic sea-level rise and subsidence is narrow, indicate that even there, the accelerated eustatic rise is an important control in the historical increase in land loss. As we will see, most of the Gulf Coast is experiencing coastal erosion and inundation that is unprecedented in historical times. This begs the question, what is causing the accelerating rate of sea-level rise this century, and how much will the rate of rise increase in the future?

The historical acceleration of sea-level rise in the Gulf of Mexico corresponds to an increase in surface water temperatures in the Gulf of Mexico this century (fig. 2.6), contributing to the increase in the rate of rise. However, there is also evidence for an increase in the meltwater contribution from glaciers (fig. 2.7). Based on the 2021 IPCC *Sixth Assessment Report*, sea level could rise as much as seven feet by the end of this century, with most of that water coming from melting ice sheets (fig. 2.21). Greenland is

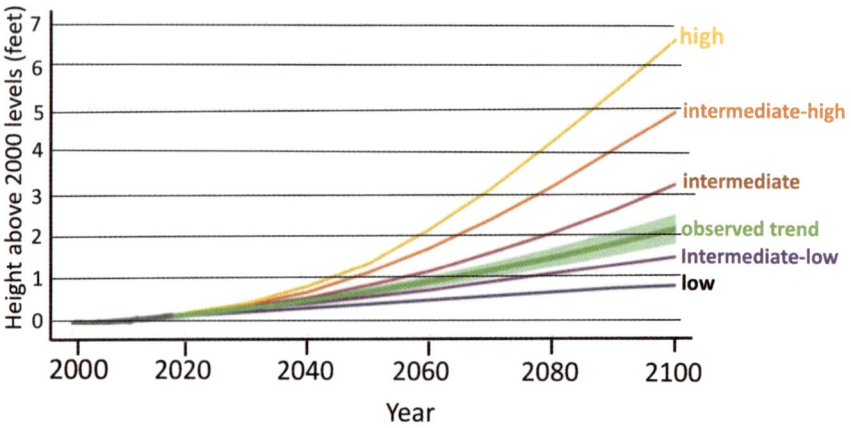

FIGURE 2.21. Observed post-2000 sea-level rise and IPCC sea-level rise predictions through 2100 based on future rates of greenhouse gas emissions, associated global warming, and potential glacier and ice sheet contributions. Modified from NOAA Climate.gov graph.

expected to be a significant contributor, but there is growing consensus that Antarctica could have the most significant contribution. The scenarios shown in figure 2.21 arise from uncertainties not so much about the magnitude of sea-level rise as about greenhouse gas emissions. This rate could be reduced by about one-third if greenhouse gas emissions were significantly reduced.

The melting of glaciers and ice sheets, the most significant contributor to global sea-level rise, began thousands of years ago. However, the recent acceleration in glacial meltwater input to the ocean far exceeds prehistoric rates. This meltwater input is mainly irreversible, short of Earth shifting back into another glaciation, which would occur at geological timescales. While surface water temperatures in the oceans increased sharply during recent decades, reversing the steric influence on sea-level rise would require centuries to thousands of years because the oceans will be slow to dissipate that heat. The same is valid for subsidence. The factors that drive subsidence, including tectonic movement, sediment loading and compaction, and glacial isostatic processes, operate at centennial to millennial timescales and are irreversible at shorter timescales. It is possible to decrease subsidence resulting from subsurface fluid withdrawal, but what is lost in land elevation is not recoverable. Thus, the historical acceleration of sea-level rise is mainly irreversible.

The epicenter of coastal inundation is coastal Louisiana. However, the leading cause of this inundation is a reduction in the sediment supply needed to compensate for the accelerating rise in sea levels. Efforts are underway to combat this problem, but most of Louisiana's coastal wetlands will be lost within a few decades. The best we can do to decrease the rate of global sea-level rise is to reduce greenhouse gas emissions and slow the rate of global warming.

Driven by uncertainties about Antarctica's contribution to sea-level rise, a group of scientists met during two NASA workshops in 1990, hosted by Dr. Robert Bindschadler of NASA. The objective was to discuss the need for a coordinated effort to investigate the interactions between ice, ocean, atmosphere, and land in Antarctica and their implications for sea-level rise. These workshops led to the creation of the West Antarctic Ice Sheet (WAIS) Initiative. I was fortunate to be one of the participants in these meetings, and I recall that our first choice of an acronym was SEARISE. That was abandoned based on arguments by a representative from the National Science Foundation, who provided funding for the workshops, that SEARISE would be too alarmist. Hence, the group adopted the name WAIS.

By the end of the first WAIS meeting, there was consensus that given the potential contribution of Antarctica to sea-level rise, there was a dire need to gather more scientists to study the problem. This was at the early stages of the satellite era of polar observation, when the scientific community was debating the reliability of tide-gauge records and evidence for the acceleration of sea-level rise. The WAIS community has grown considerably, and its annual meeting focuses on the Pine Island Bay drainage system. The scientific community has also become bolder, having adopted names like "Doomsday Glacier" that have become popular with the press. The evidence for accelerated sea-level rise in historical time has become indisputable, and field and satellite observations have yielded records of rapid change in Earth's ice sheets.

3

Climate Change Impacts

In addition to sea-level rise, impacts of climate change on the Gulf Coast include higher atmospheric and oceanic temperatures and changes in precipitation and runoff. Atmospheric and water temperatures influence water quality, coastal ecosystems, and storm intensity. Changes in precipitation and runoff can significantly alter bay and lagoon salinities and sediment supply from rivers. This chapter focuses on how increasing atmospheric temperatures and changes in precipitation have influenced the coast and how they are expected to impact the Gulf Coast in the coming decades. Background map from Google Earth.

By ~4,000 years ago, the average rate of sea-level rise in the northern Gulf of Mexico had decreased from 16 inches/century in the early Holocene to 2.0 inches/century, which should have resulted in coastal stability. However, significant changes continued to impact coastal environments at different times across the region, suggesting that regional variations in climate caused them. Parts of the Gulf Coast are experiencing climatic changes that exceed those that occurred during the late Holocene, and the impacts are becoming more evident.

The Gulf Coast spans multiple climate zones, being mainly humid from Louisiana to Florida, semiarid in east and central Texas, and arid in south Texas, with mean annual rainfall ranging from 8 to 79 inches/year (fig. 3.1). The strongest gradient occurs in Texas, where mean annual precipitation ranges from 45 inches/year at the Texas-Louisiana border to 8 inches/year near the border with Mexico. The eastern part of the Gulf Coast, from Louisiana to west Florida, is humid, with current mean annual precipitation exceeding 50 inches/year.

Atmospheric temperatures in the United States have increased on average ~1.7°C since 1895, with most of that increase occurring in the past 50 years. For those of us living on the Gulf Coast, this increase has been noticeable mainly in the form of more extreme heat waves, which have been on the rise in recent years. Most of that heat, approximately 90 percent, is absorbed by the ocean, and the Gulf of Mexico has proven to be especially vulnerable to this warming trend. Surface water temperatures have increased across the Gulf during historical time, with those in 2023 being the highest on record (fig. 2.6).

Interannual to interdecadal meteorological cycles influence variations in Gulf Coast temperatures and precipitation. Most influential among these are three-to-seven-year El Niño–Southern Oscillation (ENSO) cycles in temperature and precipitation. These cycles are driven by sea surface temperatures in the eastern tropical Pacific Ocean and transmitted across the southwestern United States. El Niño events bring lower winter and spring temperatures and greater annual rainfall. They alternate with La Niña events, which are characterized by warmer temperatures and drier conditions, including regional droughts. Climate models indicate that these alternating conditions result in significant changes in coastal circulation driven by changes in salinity and winds. El Niño and La Niña events are expected to intensify in the future, likely resulting in more prolonged dry spells (droughts) and wet spells (floods).

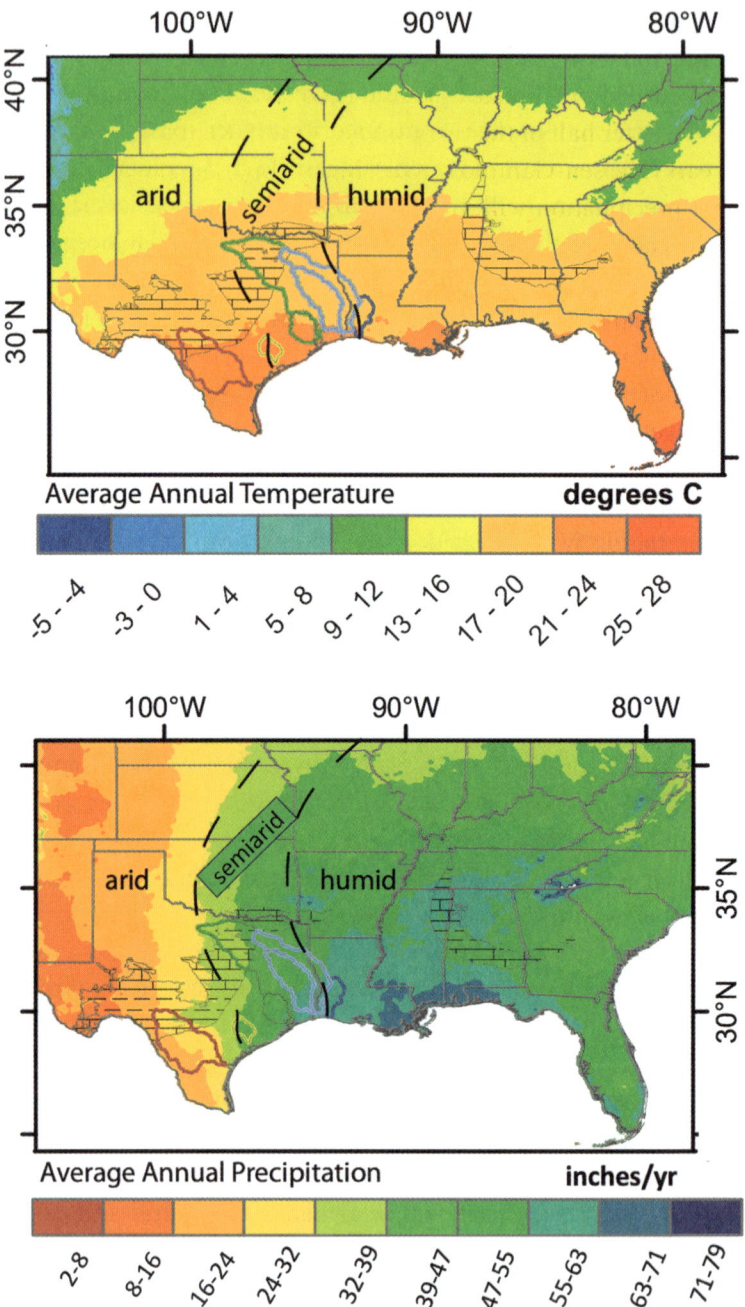

FIGURE 3.1. Average annual temperatures and precipitation for the Gulf Coast. Modified from Milliken et al. (2017). Reprinted with permission from SEPM.

Areas most prone to the impacts of climate change are those with the largest precipitation gradients, with west Texas being especially vulnerable to increased aridity. Climate models indicate drier conditions in the region during the latter half of the twenty-first century than during the previous 1,000 years (Nielsen-Gammon, 2021). Meanwhile, climate models also suggest that precipitation will increase across the southeastern states in the coming decades. The most significant increase will occur along the coast, mainly because of warmer Gulf temperatures and increased evaporation. Average annual precipitation has already increased 20–30 percent between the Florida Panhandle and Louisiana.

An important impact of changing precipitation will be changes in coastal vegetation, which can influence erosion rates and sediment discharge of rivers and streams and alter wind erosion and transport. Currently, the Gulf Coast has distinct vegetation zones, with the eastern Gulf Coast dominated by temperate evergreen forests, south Florida supporting more tropical vegetation, and most of the Texas coast inhabited by live oak forests. Results from numerical modeling aimed at testing the tolerance of Gulf Coast forests to climate change show south Texas experiencing the most dramatic change with the northward expansion of tropical dry forests, similar to those of the east coast of Mexico (Liénard et al., 2016). More arid conditions will likely result in an expansion of dune fields that remove sand from the coastal system. In contrast, the eastern Gulf Coast is expected to experience only minor changes in coastal vegetation, except for the southwest Florida coast, where rising sea level is expected to significantly alter coastal vegetation in the Everglades.

Paleoclimate Records

Historical records of temperature and precipitation for the Gulf Coast region typically span just over a century, and satellite imagery needed for analysis of atmospheric circulation spans only about half a century. These records are too short to allow studies of decadal-scale climate variability. Paleoclimate records can help fill this gap and, combined with improved aerial imagery and results from field-based observations, allow comparison of Gulf Coast climate conditions to changes in coastal environments. Several paleoclimate proxies are available to achieve this objective, each providing information about past climate conditions.

Oxygen isotope records from ice cores and cave deposits yield pale-otemperature records with a precision of tenths of a degree Celsius and provide decadal-to-century-scale resolution that spans thousands of years. These data provide a record of global temperature change that shows significant warming between ~11 ka and ~10 ka as the Earth shifted out of its late Pleistocene icehouse condition to the generally warmer conditions of the Holocene (fig. 3.2). This warming trend lasted through the *Medieval Warm Period* and ended with the Little Ice Age, which occurred between ~600 and 200 years ago. Since the Little Ice Age, global temperatures have increased rapidly in response to human climate alteration. This dramatic shift to warmer conditions was the focus of a seminal paper published in 1998 by Dr. Michael Mann and colleagues, which introduced what was subsequently referred to as the "hockey stick" curve. When this curve was published, it spurred debate about the magnitude and significance of temperature change. Since 1998, temperatures have continued to increase, so the stick has gotten longer, and the debate has diminished.

Good regional records of temperature and humidity can be obtained from fossil pollen and spore assemblages using sediment cores from lakes, lagoons, estuaries, and wetlands. These aquatic settings typically contain organic-rich sediments that favor the preservation of this fossil material and yield reliable radiocarbon ages. Another method involves the analysis of tree rings, which relies on tree ring widths and density to estimate precipitation. Trees in temperate regions have a distinct growing season. Each ring records one year, and some trees yield annual resolutions spanning hundreds of years. Scorched tree rings record wildfires, which are usually associated with droughts. These methods have been applied sparingly to reconstruct late Holocene climate conditions in the Gulf Coast region. However, the results have been very informative about how the warmer future may impact the coast.

Regarding changes in precipitation, Texas has the greatest precipitation gradient and is expected to undergo the most significant change in the future. So it is no surprise that paleoclimate studies have revealed significant spatial and temporal (millennial to centennial timescale) variability in temperature and precipitation across Texas during the late Holocene. For example, during the late Holocene, Baffin Bay shifted dramatically from more "normal" estuarine salinities to hypersaline conditions. This change was caused partly by the increased isolation of the bay by landward migra-

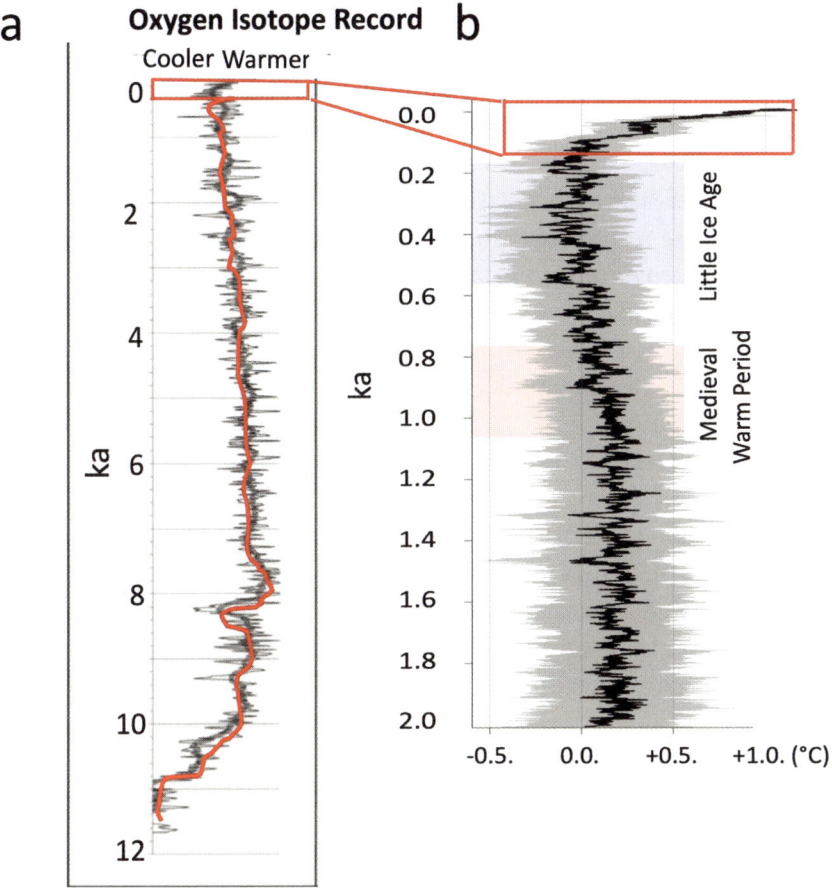

FIGURE 3.2. (a) Oxygen isotope records show how the climate evolved over thousands of years as the Earth transitioned from its icehouse state to warmer conditions, with the most pronounced change occurring between 11 ka and 10 ka. (b) Expanded isotope record for the past 2,000 years (based on Marcott et al., 2013 and Rasmussen et al., 2006). The red box denotes the rapid temperature shift commonly called the "hockey stick."

tion of Padre Island. Still, the timing and rate of this change indicate that it was mainly a response to increasing aridity in south Texas during this time (Simms et al., 2010). To the north, the Nueces River bayhead delta in Corpus Christi Bay shifted location several times, linked to alternating wet and dry conditions over centuries (Rice et al., 2020). Associated with these changes was a northward shift in oyster reefs as bay salinities gradually increased (Goff et al., 2016). Farther north, pollen records from Galveston Bay, located within the semiarid climate zone, show that the estuary has experienced increasingly humid conditions during the past 3,000 years (Ferguson et al., 2018b). Thus, the paleoclimate record for the Texas coast is one of increased aridity in south Texas, variable conditions in central Texas, and increasing precipitation in east Texas.

Paleoclimate records remain sparse for the eastern Gulf Coast. A few well-dated pollen records from lakes in Alabama and Florida show a change from oak-dominated forests in the early Holocene to pine-dominated forests in the late Holocene. This change indicates a shift toward warmer winters and increased humidity during the past several thousand years.

Sediment Delivery to the Coast by Rivers

One of the main impacts of climate change on coastal areas is alterations in sediment supply from rivers and streams, caused mainly by variation in precipitation and associated changes in river discharge. Sediment supply from rivers is also controlled by the size, relief, geology, and vegetation cover of a river's drainage basin. Drainage basin size, relief, and geology do not vary significantly over decades to centuries, but precipitation greatly influences sediment discharge at these timescales (Anderson et al., 2022). Vegetation cover is also correlated to precipitation. Generally speaking, decreasing vegetation cover results in greater erosion and sediment discharge. Historically, humans have assumed a significant role in river sediment discharge by converting forests to agricultural land. The result was increased sediment discharge to estuaries and the Gulf during the 1800s and early 1900s. This was followed by the construction of dams and flood control structures, and alteration of river drainage systems that reduced sediment discharge.

A total of 33 rivers deliver sediment to the US Gulf Coast. These rivers vary widely in terms of their drainage basin size and climate setting, particularly their rates and timing of rainfall, which equates to significant differences in freshwater and sediment discharge to the coast (fig. 3.3). The Mississippi-Atchafalaya watershed is just over 1,245,000 square miles in area and accounts for approximately 82 percent of the total freshwater discharge, and by far the largest sediment discharge (427.90×10^6 tons/yr). It has also been the dominant sediment source in the Gulf of Mexico Basin

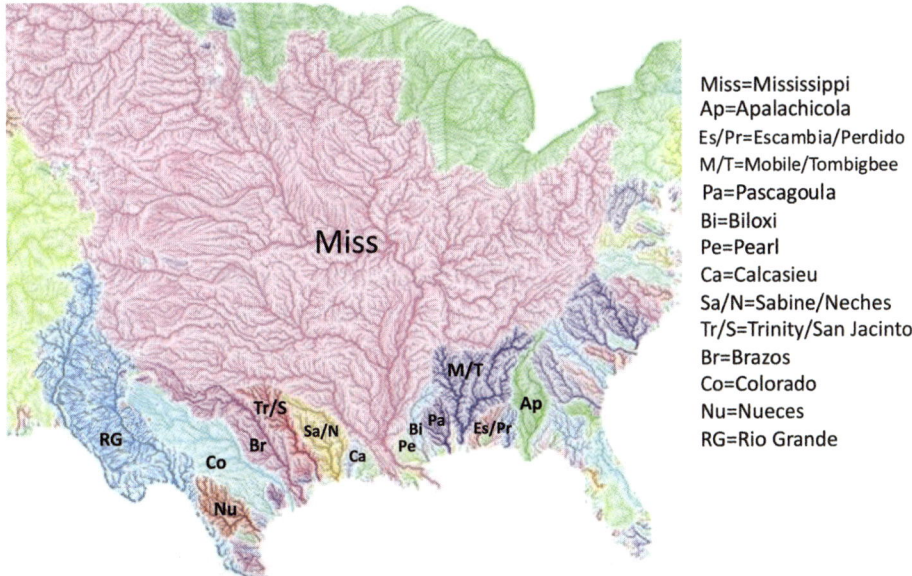

Miss=Mississippi
Ap=Apalachicola
Es/Pr=Escambia/Perdido
M/T=Mobile/Tombigbee
Pa=Pascagoula
Bi=Biloxi
Pe=Pearl
Ca=Calcasieu
Sa/N=Sabine/Neches
Tr/S=Trinity/San Jacinto
Br=Brazos
Co=Colorado
Nu=Nueces
RG=Rio Grande

FIGURE 3.3. Drainage basins of rivers flowing into the Gulf Coast. Modified from Robert Szucs.

for millions of years, and it remains the dominant source of fine-grained sediments for the Mississippi, Louisiana, and Texas continental shelves. Other large rivers, such as the Apalachicola, Brazos, Alabama-Tombigbee, Colorado, and Rio Grande, formed large deltas on the continental shelf in the past when climate conditions were more humid in their drainage basins (fig. 1.5). These deltas retreated landward when the rate of sea-level rise was decreasing, but in west Texas precipitation was also decreasing, so changes in climate played a vital role in the demise of the Brazos, Colorado, and Rio Grande deltas (Anderson et al., 2022).

Most rivers of the Gulf Coast have relatively small drainage basins (less than 37,000 square miles). During the early Holocene, their river valleys were drowned to form estuaries. So they never formed large deltas on the continental shelf, but they did construct bayhead deltas that are strongly influenced by sea-level rise and changing climate, mainly in the western Gulf Coast region.

By far, the most significant portion of sediment delivered to the Gulf of Mexico by the Mississippi River is fine-grained material that is transported in suspension to the coast and then to the west by surface currents (fig. 3.4). As a result, the coastal waters of western Louisiana and Texas are laden with sediment. This sediment dispersal system has been active for most of Holocene time, resulting in a thick layer of mud on the central and south Texas continental shelf, known as the Texas Mud Blanket. Available evidence points to climate-induced changes in ocean circulation patterns as the cause of increased mud blanket growth during the past 4,000 years (Weight et al., 2011). This mud blanket expansion resulted in the burial of deltas on the central and south Texas continental shelf and the burial of coral reefs that were once widespread along the central Texas coast.

Rivers flowing into the eastern Gulf deliver relatively small amounts of fine-grained sediment to the coast. This results in generally clear coastal waters for Florida and Alabama, in contrast to the turbid coastal waters

FIGURE 3.4. Satellite image shows how sediment-laden surface waters that emanate from the Mississippi River are transported hundreds of miles west by wind-driven currents. Over thousands of years, this process has deposited a thick layer of mud on the western Louisiana and Texas continental shelves and nearshore areas. It accounts for the western Gulf's generally turbid coastal waters relative to the eastern Gulf's clearer waters. NASA image.

of the western Gulf. But these differences are not just aesthetic; they also strongly influence coastal ecosystems, especially coastal wetlands that require a steady sediment supply to keep pace with rising sea levels.

Texas has multiple rivers that vary significantly in terms of drainage basin area and climate setting (fig. 3.3). This has resulted in significant differences in the long-term sediment delivery rates from these rivers to the coast. The largest are the Brazos, Colorado, and Rio Grande. The Brazos and Colorado Rivers have drainage basins that extend into west Texas and New Mexico. The Rio Grande's drainage basin extends across Texas and New Mexico into Colorado. As a result, these rivers span a wide precipitation gradient that makes them sensitive to climate change. The Rio Grande has the second largest drainage basin of any Gulf Coast river (500,200 square miles), but a sediment discharge of only 36.9×10^6 tons/year. Its disproportionate drainage basin area and sediment discharge ratio relative to the Mississippi River is a product of its more arid climate setting; much of the Rio Grande drainage basin includes ancillary basins that yield little sediment to the Rio Grande during dry periods. The river is also experiencing excessive agricultural and domestic water usage.

Relief of a river's drainage basin is another important control on sediment discharge. The topography of the *coastal plain*, the relatively flat, low-lying land adjacent to the coast, is significant. The coastal plains of Mississippi, Alabama, and Florida are somewhat narrow and steep, with rivers flowing through narrow branching valleys that extend within miles of the coast. These include the Apalachicola River, which has the largest drainage basin in Florida (19,500 square miles), and the Blackwater, Escambia, and Perdido Rivers. These rivers occupy drainage basins with extensive Pliocene and Pleistocene sandy strata, which have been important sources of sand for the west Florida and Alabama coasts. The Suwannee, Econfina, St. Marks, and Ochlockonee Rivers all drain into the Florida Big Bend area. These rivers have relatively small drainage basins and flow across regions of low relief etched into carbonate rocks, which accounts for their low water and sediment discharge. The Mobile-Tombigbee drainage system occupies a deeply incised valley within a humid climate zone. Its sediment supply to the Mobile bayhead delta has varied little during the late Holocene and is not expected to change significantly in the coming decades (Rodriguez et al., 2008). Likewise, the Pascagoula and Pearl Rivers, which drain into Mississippi Sound, are not expected to have their sediment discharge significantly influenced by climate change.

The coastal plains of Louisiana and Texas extend tens of miles inland. They have relatively low relief and broad, meandering river channels extending to the coast. The lower reaches of the drainage basins of these rivers are occupied chiefly by fine-grained sedimentary strata, resulting in large suspended loads. Sand delivered to the coast by these rivers originates mainly from the erosion of older river channels as the rivers meander across the landscape. Sediment discharge in Texas rivers is expected to change significantly in response to climate change.

The Impact of Increasing Aridity

The Mississippi River drainage basin spans multiple climate zones. It extends far enough north to have been strongly influenced by glacial outwash, which accounts for large quantities of the sediment delivered to the Gulf of Mexico Basin during the late Pleistocene and Holocene. During historical times, sediment delivery from the Mississippi to the coast has been drastically reduced by alterations in the natural water and sediment dispersal system, including the construction of levees and dredging of irrigation and floodwater channels. There is also evidence that climate change is starting to influence the Mississippi River's discharge. In the summer of 2023, abnormally hot and dry conditions led to an unusual drop in water levels that threatened navigation in the upper reaches of the river channel. At the same time, the incursion of salty Gulf water (saltwater wedge) threatened drinking water supplies in New Orleans. This saltwater incursion was exacerbated by the deepening of the channel from 45 to 50 feet to allow larger ships to transport cargo to Louisiana ports.

The Brazos, Colorado, and Rio Grande are situated mainly within west Texas and New Mexico and have been significantly influenced by the region's alternating humid and arid conditions. Chapter 4 describes how changes in sediment discharge from these rivers resulted in significant shifts in their deltas and, ultimately, sediment supply to the coast during the late Holocene. Climate models call for increasing aridity in the southwestern United States that, coupled with human alteration of river discharge and sediment supply to the coast, will likely equal or exceed changes that occurred during the late Holocene. Currently, the Colorado River is struggling to maintain its bayhead delta. The Rio Grande's sediment discharge has been severely impacted by increasing aridity and human influence, and it is no longer a viable source of sediment to the coast.

A study of the Brazos River delta has revealed a historical record of episodic flow and sediment transport driven by strong precipitation events (Rodriguez et al., 2000). Sediment discharge is generally low during warmer, drier summer months associated with La Niña events. High-discharge events occur mainly during El Niño years when moisture from the Pacific Ocean is transported east across the southwestern United States. These high-discharge events remobilize sand within the river channel and deliver it to the coast, resulting in episodic growth of the Brazos delta. Photographic records show that the delta's growth has decreased in recent decades, a response to increasingly drier conditions in the Southwest.

Other Texas rivers, including the Sabine, Trinity, San Jacinto, Lavaca, San Antonio, and Nueces, have much smaller drainage basins than the Brazos, Colorado, and Rio Grande. This equates to lower water and sediment discharge for these rivers. They also span fewer climate zones, so they tend to experience more consistent precipitation and discharge, with discharge rates generally decreasing from east to west. Because significant portions of their drainage basins are within 100 miles of the coast, high-discharge events for these smaller rivers are commonly associated with tropical disturbances in the Gulf of Mexico.

Unlike the valleys of the Brazos, Colorado, and Rio Grande, the valleys of smaller Texas rivers were flooded during sea-level rise of the last ~10,000 years to form estuaries. The rivers flowing into these estuaries have formed bayhead deltas that struggle to keep pace with rising sea levels. Studies have shown that changes in sediment discharge from these smaller rivers during the late Holocene were regulated by climate, particularly precipitation (Anderson et al., 2022). A good example is the Trinity bayhead delta in Galveston Bay. Between 2.6 ka and 1.6 ka, the delta grew significantly. Pollen records from the delta indicate that this was a time when the local climate was transitioning from more arid to more humid conditions (Ferguson et al., 2018a). During this same time, both the Sabine and Calcasieu bayhead deltas shifted landward approximately nine miles, while the Lavaca bayhead delta did not change significantly (Milliken et al., 2008b, c; Maddox et al., 2008). These observations reveal an out-of-phase shift toward more arid conditions in south Texas and more humid conditions in east Texas, similar to the historical trend. Climate records from 1895 through 2020 show increased precipitation in east Texas, while the western half of the state has had decreased precipitation. Climate models indicate this trend will continue in the coming decades (Nielsen-Gammon et al., 2020). This suggests that sediment supply to the Texas coast from the Brazos, Colorado, and Rio Grande will continue to decrease. Sand supply from the Rio Grande and Colorado is already minimal, so the increase needed to help sustain the Matagorda Peninsula and South Padre Island is not expected. The impact of climate change on the Brazos River is uncertain, but climate models provide little encouragement for an increase in its sediment supply to the coast. The Nueces bayhead delta is the largest bayhead delta on the central Texas coast and a vital component of the Corpus Christi Bay and Nueces Bay ecosystem. Drier conditions are expected to result in significant landward retreat of the delta in the next few decades as the rate of sea-level rise increases and sediment supply to the delta decreases.

The Trinity delta in Galveston Bay is the largest bayhead delta on the east Texas coast. Its low-elevation delta plain is highly vulnerable to sea-level rise. Likewise, the low-lying coastal plains adjacent to Sabine Lake and Calcasieu Lake depend on a steady river sediment supply to combat rising sea levels. These areas are undergoing rapid rates of coastal inundation that are projected to increase because sediment supply from area rivers is not keeping pace with sea-level rise (Milliken et al., 2008b, c).

The projections for decreasing sediment delivery to the coast are alarming, given the lasting impacts of human alteration of river channel networks and dam construction. In general, agricultural activity within most Gulf Coast drainage basins began over a century ago and resulted in increased sediment supply to streams and rivers. Dams came later and have reduced sediment supply, mainly sand, which tends to be trapped in reservoirs upstream of dams. The Brazos River basin includes 13 dams and associated reservoirs that trap sand delivered principally from the upper reaches of the drainage basin. Sand that is now delivered to the coast by the river comes primarily from the erosion of bank deposits in the lower reaches of the basin.

Another concern about river discharge is nutrient loading in response to increased agricultural activity, which can result in the depletion of oxygen, known as hypoxia, within estuaries and offshore waters. The most notable example is the formation of oxygen-depleted bottom waters offshore of the Mississippi Delta during increased freshwater outflow. The decay of suspended organic material results in oxygen depletion, commonly producing a "dead zone" where animals cannot survive. The Union of Concerned

Scientists has concluded that the economic impact of dead zone expansion has been as much as $2.4 billion per year since 1980.

The extent of this dead zone increases during spring runoff when discharge is high and water temperatures increase. Research has shown that this process has occurred for at least a thousand years. Still, increased agricultural activity in the river basin and associated increased nutrient flux have exacerbated the impacts (Osterman et al., 2009). These events are expected to occur more frequently in coming decades, and the size of the dead zone will likely increase as Gulf water temperatures rise. Similar impacts will likely occur in estuaries and, although more minor in scale, will have widespread ecological effects. One example is Tampa Bay, where increased organic sediments from storm and industrial drainage have led to oxygen depletion and associated fish kills.

Extreme Weather Events

The past two decades have been a time of extreme weather events in the Gulf Coast region, including heat waves, cold waves, floods, and droughts with little historical precedent. The magnitude of floods appears to be increasing, and the number of extreme heat waves impacting the United States has increased from two per year to six per year since the 1960s. Record-breaking high temperatures persisted across the Gulf Coast during the summer of 2023. In 2021, Texas was hit by an unprecedented winter storm with temperatures well below freezing for nine days, resulting in the failure of the state's power grid and loss of lives, making it one of the costliest weather disasters in Texas history.

Scientists have argued for decades that high latitudes are more sensitive to global warming than lower latitudes. Still, the extent and magnitude of their influence on weather patterns in lower latitudes have been underestimated. The Arctic has been warming twice as fast as the rest of the planet since 1995. One significant result has been a decrease in the extent of sea ice in the Arctic Ocean. The extent of summer sea ice cover in the Arctic has shrunk by 13 percent per decade since 1979 and is currently at its lowest area on record. Less extensive sea ice means more open water to absorb the sun's energy, known as the *albedo effect*, which leads to even warmer air masses and ocean surface waters. Sea ice cover in the Arctic is projected to be virtually gone by 2050. Record low sea-ice coverage in Antarctica during the 2023 austral summer suggests that the Southern Hemisphere may follow a similar pattern.

Another critical effect of warmer conditions in the Arctic has been a weakening of the jet stream, a west-to-east-flowing air mass between cold Arctic air and warmer air to the south. As the Arctic becomes warmer, the temperature gradient decreases, and the jet stream's speed decreases. This causes it to meander more and dip farther south (fig. 3.5). Dips in the jet stream have strongly influenced the intensity and path of storms, prolonged extreme heat waves, and allowed cold Arctic air masses to extend as far south as the Gulf Coast. They have also contributed to more frequent summer heat waves and droughts, increased the magnitude and duration of floods, and altered the paths of severe storms. The 2021 Texas Freeze was associated with a dip in the jet stream that led to the penetration of subzero Arctic air into south Texas. A similar but less extreme event occurred in 2024. Some climate models indicate that increasing temperatures will exacerbate this *Arctic amplification* and thus intensify extreme weather events around the globe, including the Gulf Coast region. More recently, scientists have learned that declining sea ice in the Arctic is exacerbated by atmospheric rivers, which have been increasing in response to anthropogenic warming (Zhang et al., 2023). Exactly how these extreme weather events will impact the Gulf Coast remains uncertain, but they will likely amplify drought conditions in the western Gulf Coast region. They may also drive these drought conditions farther east and thus impact vegetation and aquatic life.

Oceanographic Changes

Oceanographic circulation in the northern Gulf of Mexico is driven predominantly by winds, with tides being a secondary influence. Wind- and wave-driven current patterns influence sand transport along the coast, and oceanographic circulation on the continental shelf strongly influences the dispersal of fine-grained suspended sediments and nutrients. The

FIGURE 3.5. *Facing*, (a) Map of the United States showing the sinuous path of the jet stream resulting in southern penetration of cold Arctic air in the west during winter months. (b) The same pattern of increased sinuosity of the jet stream can result in unseasonably warm conditions in the eastern United States during warmer months. Modified from WeatherBELL.com.

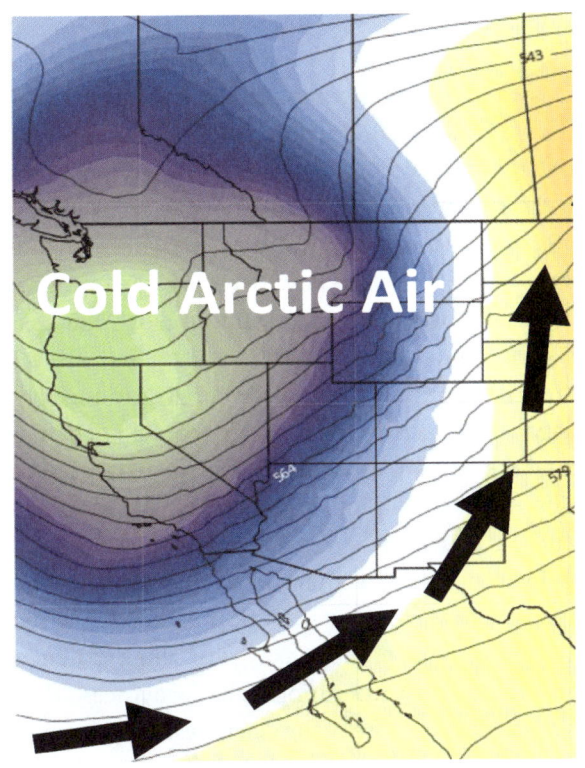

a

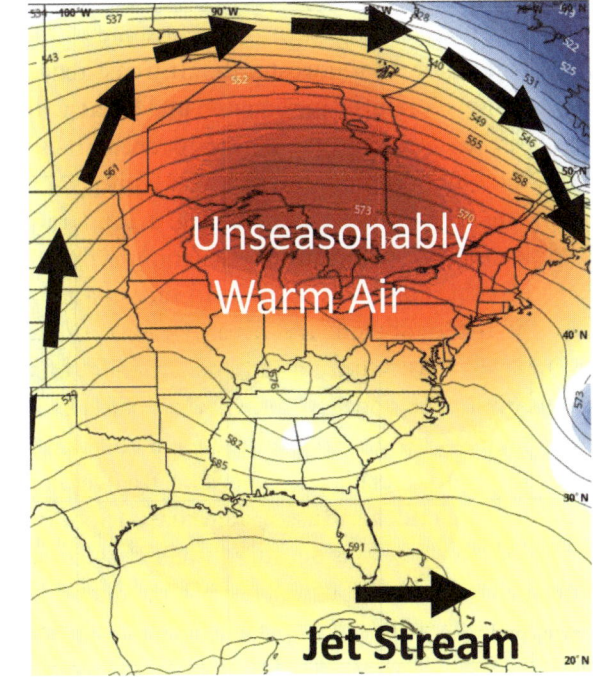

b

direction of coastal currents is controlled by the shape and orientation of the coast and by prevailing wind patterns. One of the biggest uncertainties in predicting the impacts of climate change concerns potential changes in these wind patterns and associated oceanographic changes, which could alter longshore and wind-driven currents. Differences in the direction and intensity of longshore circulation have been considered a possible cause of observed changes in coastal barrier evolution along the Texas coast over the past several thousand years (Anderson et al., 2022). Still, it takes work to isolate this effect from other factors that regulate sand supply to the coast.

The intensity and direction of surface currents in the Gulf of Mexico vary throughout the year, mainly because of seasonal temperature changes. There is also significant variation in sea surface temperatures in the Gulf, which influences circulation by changing water column stratification. West of the Mississippi River Delta, surface currents flow mainly toward the west, while east of the delta circulation is more complex, in part because of the irregular shape of the coast. The Gulf Loop Current marks the boundary between these two oceanographic provinces. This surface current transports warm waters from the Caribbean Sea through the Yucatán Channel and into the eastern Gulf of Mexico (fig. 2.20). The behavior of the Loop Current varies seasonally in response to migration of the *Intertropical Convergence Zone* (ITCZ). During warmer months, the ITCZ shifts north, forcing the Loop Current farther north in the Gulf of Mexico. During the winter, the ITCZ moves south near the equator, prevailing winds shift from southeasterly to northeasterly, and the Loop Current shifts south.

About every 3 to 17 months, counterclockwise eddies spin off the Loop Current and migrate into the western Gulf (fig. 2.20). Landward migration of these eddies results in sea-surface elevation changes of up to two feet in the central Gulf. Along the south Florida coast, astronomical tides are exacerbated by meteorological and oceanographic tides, resulting in "king tides." In the western Gulf, an anticipated effect of warmer Gulf temperatures and more prolonged summer wind patterns is more landward migration of eddies, leading to more frequent and greater-amplitude oceanographic tides. South Florida has experienced a fivefold to twelvefold increase in extremely high tides and associated coastal flooding since 2000, a 75 percent increase over previous decades (Sweet et al., 2019).

Evidence shows that the Loop Current has been weakening this century, especially during the past two decades, as surface water temperatures have increased (Piecuch, 2020). The expected impact of this weakening is more frequent high-tide flooding. A recent NOAA report concludes that by 2050, high-tide flooding in the eastern Gulf will occur on average about 25 to 80 days per year. The same NOAA report predicts that high-tide flooding in the western Gulf could occur 80 to 185 days a year by 2050. These predictions are difficult because time-series oceanographic data are too sparse to relate tidal changes to oceanographic circulation patterns. Soon, new-generation satellite data will fill this gap and significantly improve our understanding of oceanographic tides and their cumulative impacts on coastal environments.

I used to instruct students not to confuse weather and climate because they operate at different timescales. More recently, I have come to recognize that extreme weather events are, more often than not, products of climate change. This is because changes in our climate have occurred so rapidly that it is exerting a clearer influence on weather patterns. Rapid intensification of hurricanes caused by warmer Gulf of Mexico waters and extreme weather events caused by Arctic amplification are examples. Indeed, the impacts of climate change are occurring so fast that the rate of change is outpacing our understanding of the climate system and our ability to model the interactions between climate, the oceans, and other global influences. So it should come as no surprise that climate change is causing coastal impacts that were not predicted and, for that matter, still need to be better understood. I call this the climate domino effect, where changes occurring in distant areas impact our local environment. A good example is the linkage between the Gulf of Mexico's weather and climate patterns with changes taking place in the Arctic.

4

Demise of Gulf Coast Deltas

Most of the world's great deltas grew during the late Holocene as the rate of sea-level rise decreased. This century's acceleration of sea-level rise has tipped the scales against their sustainability. Humans have further contributed to their demise by damming rivers and altering their sediment delivery and dispersal systems. Background map from NASA.

Deltas form where the sediment supply is large enough to compensate for relative sea-level rise (RSLR) and the day-to-day influence of wave erosion. If sediment supply to the delta is high, the river will construct lobes of sediment that advance seaward beyond the zone of wave reworking to construct a fluvial-dominated delta made up of a subaerial *delta plain* and submarine delta (fig. 1.1). If sediment supply is low, sediments are reworked by waves to form a wave-dominated delta that is unable to advance seaward of the coastal zone, resulting in a *delta headland*. If the river flows into an estuary, it forms a bayhead delta.

The modern marine deltas of the world initially formed 8.5 ka to 6.5 ka as the rate of global sea-level rise decreased. Many of them are archaeological sites that date back at least 7,000 years, indicating that they were prime locations for human occupation. Today, they remain locations of some of the largest human population sites and include major centers of agriculture and commerce. They are also among the most threatened coastal environments because they depend on maintaining the delicate balance between sediment supply and rising sea levels. Most of the world's great deltas, including the Mississippi River Delta, are losing this battle (Syvitski et al., 2009). Two seminal papers published over a decade ago by Dr. John Day and colleagues in 2007 and by Dr. Michael Blum and Dr. Harry Roberts in 2009, all with Louisiana State University, noted that human alteration of sediment supply to the Mississippi River Delta and accelerated sea-level rise have rendered the delta unsustainable over the next several decades. Since then, other research has added support to this bleak scenario.

The Mississippi River Delta is not the first Gulf Coast delta to have faced demise. The Brazos, Colorado, and Rio Grande rivers all nourished large fluvial-dominated deltas during the late Pleistocene and early Holocene (fig. 1.5). Their transition from fluvial-dominated to wave-dominated deltas occurred during the late Holocene as the rate of sea-level rise was decreasing, indicating that their decline was caused primarily by reduced sediment supply. Their continued decline during historical time is the result of accelerated sea-level rise and alteration of their sediment supply and dispersal systems.

The Era of Giant Deltas

Between ~120,000 ka and ~20,000 ka, sea level fell nearly 400 feet, resulting in shoreline migration to the edge of the continental shelf. During this time, larger rivers of the Gulf Coast region experienced cooler and more humid conditions, and falling sea levels resulted in the incision of river valleys and expansion of tributary streams. The combined effect was increased sediment delivery from these rivers to the coast and the construction of deltas that advanced across the continental shelf. The coast was then occupied by vast cypress swamps that stretched hundreds of miles from east Texas to west

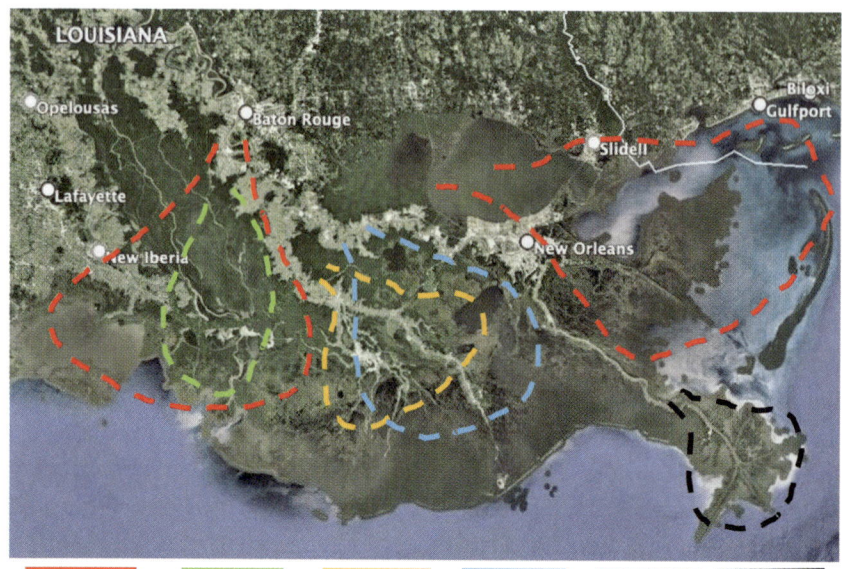

Sale-Cypremort	Atchafalaya-Wax	Teche	Lafourche	St. Bernard	Plaquemines
(4500-3500 BP)	Lake (400-Present)	(3500-2800 BP)	(1000-300 BP)	(2800-1000 BP)	(750-Present)

FIGURE 4.1. Late Holocene delta lobes of the Mississippi River Delta. Lobe ages from Yao et al. (2020). Background image from Google Earth.

Florida. Seismic data and drill cores from the western Louisiana and Texas continental shelf were used to reconstruct the evolution of these deltas as they advanced and retreated in lockstep with fluctuations in sea level that were on the order of 50 to 150 feet and had a duration of ~12,000 years for each cycle (fig. 1.4a). With each sea-level fall, the deltas advanced seaward, followed by landward retreat and erosion during subsequent rises. To the east, the Mobile, Perdido-Escambia, and Apalachicola Rivers also constructed deltas, but minimal subsidence on the continental shelf limited their preservation (Bart and Anderson, 2004; McKeown et al., 2004). Remnants of these deltas can be found on the outer continental shelf and upper continental slope (fig. 1.5). Meanwhile, higher subsidence rates in the western Gulf of Mexico Basin favored the preservation of the ancestral Mississippi, Brazos, Colorado, and Rio Grande deltas. The valleys of smaller Gulf Coast rivers were drowned to form estuaries with bayhead deltas.

By late Holocene time, the average rate of sea-level rise decreased to 2.0 inches/ century and the Mississippi River began constructing a series of large *delta lobes* offshore of Mississippi and Louisiana (fig. 4.1). With each phase of lobe formation, the river mouth advanced seaward until the channel was abandoned and a new channel was occupied, a process known as *channel avulsion*. The river has constructed six major delta lobes since ~4.6 ka. Four of these, the Sale-Cypremort, Teche, St. Bernard, and Lafourche lobes, were abandoned, while the Plaquemines and Atchafalaya deltas remain active. The abandoned lobes were reworked by waves to form fringing barrier islands surrounding the delta (see chapter 5). During the same time that the Mississippi River was growing by delta lobe construction, drier conditions in Texas resulted in lower sediment discharge rates for the Brazos, Colorado, and Rio Grande, and their fluvial-dominated deltas became wave-dominated deltas (Anderson et al., 2022). Historically, increasing aridity has led to the complete demise of the Rio Grande delta and the conversion of the Colorado delta to a relatively small bayhead delta. Only the Brazos River continues to supply enough sediment to the coast to maintain a wave-dominated delta.

The Demise of the Mississippi River Delta

The Mississippi River has by far the largest drainage basin and most significant water and sediment discharge rate of any US river, with an annual sediment load of roughly 550 million metric tons delivered to the delta and Gulf of Mexico each year (fig. 3.3). The river has constructed one of the world's largest deltas, with an area of 7,288 square miles. It is a classic "bird's foot delta" with branching distributary lobes (fig. 4.2). Additional sediment is transported to the Gulf by the Atchafalaya River through an artificial channel constructed in 1942 by the US Army Corps of Engineers to restrict river flow and reduce flooding in Morgan City, Louisiana. Sediment transported through the Atchafalaya channel has constructed the Wax Lake delta (fig. 4.3).

Most of the subaerial Mississippi River Delta and Wax Lake delta is composed of wetlands less than two feet above sea level, so these deltas are highly vulnerable to sea-level rise. According to a 2021 report issued by Louisiana's Coastal Protection and Restoration Authority, the average subsidence rate in coastal Louisiana varies between ~12 and ~30 inches/century. Given a modest prediction for eustatic sea-level rise, most of the subaerial delta, ~9,300 square miles, will be drowned by 2050 (fig. 4.4).

Efforts to slow the rate of land loss in south Louisiana will succeed only where the rate of relative sea-level rise can be balanced by sediment supply, which is pretty much dependent on the rate of eustatic rise this century and

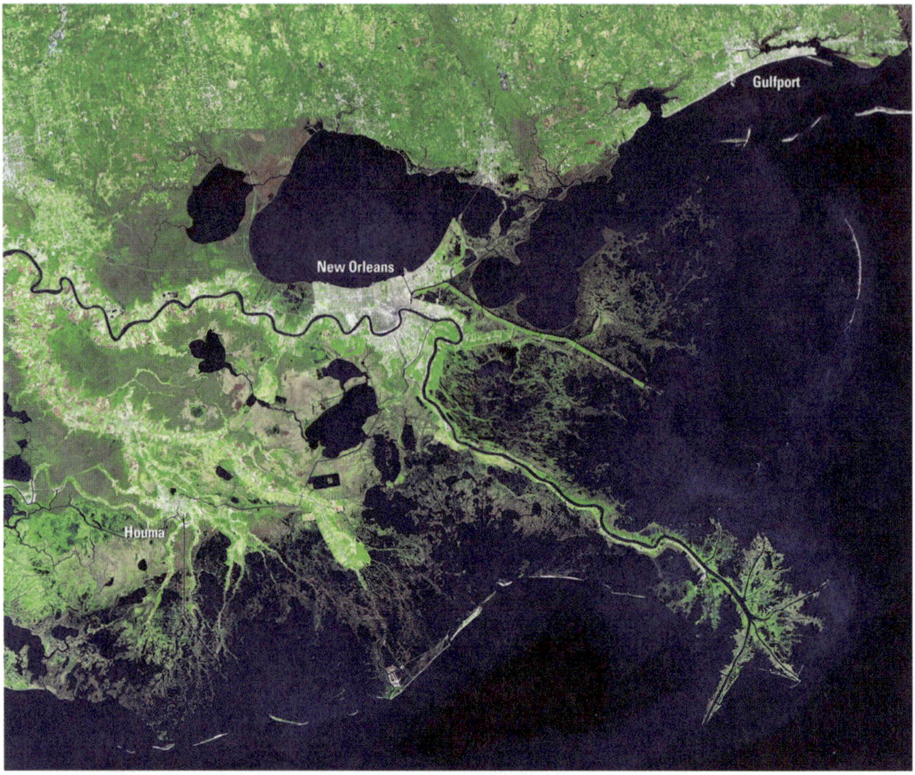

FIGURE 4.2. *Above*, The modern Mississippi River Delta as viewed from space. NASA image.

FIGURE 4.3. *Left*, The Wax Lake and Atchafalaya deltas. NASA image.

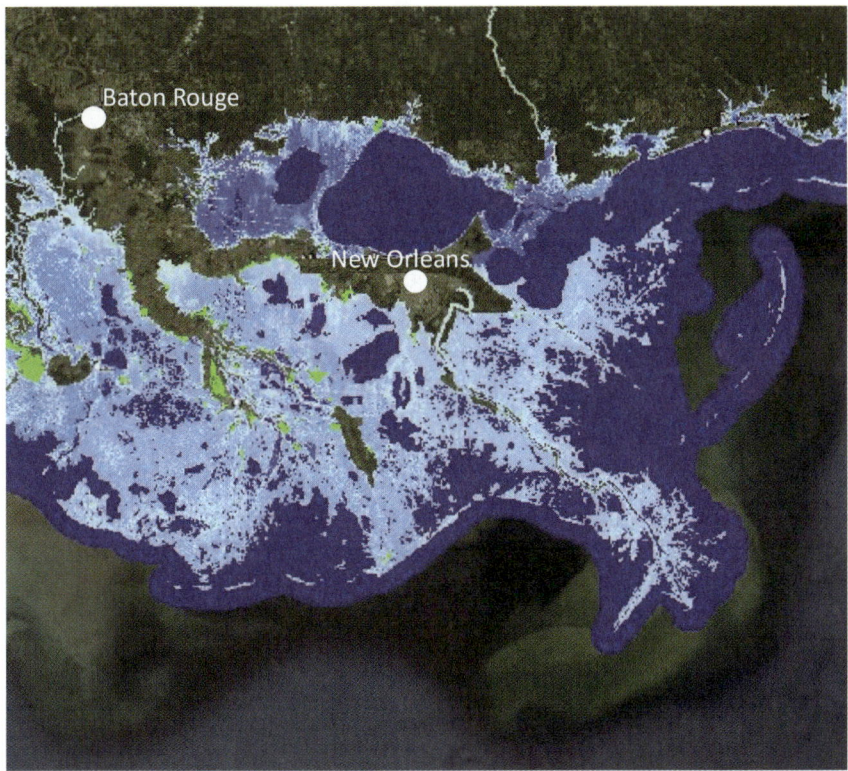

FIGURE 4.4. Model results from NOAA show portions of the Mississippi River Delta that are expected to be inundated by 2050, shown in light blue. Dark blue designates areas currently below sea level.

rates of subsidence. Subsidence varies widely across the region (fig. 2.16). Research has yielded compelling evidence that sediment compaction is the leading cause of regional subsidence in south Louisiana. This is common for deltas because they are regions of rapid sediment accumulation, generally equating to greater thicknesses of water-saturated, loosely compacted sediments. In south Louisiana, it is estimated that 60–85 percent of the total subsidence results from the compaction of sediments within the upper 15 to 30 feet of the sediment column, where more compressible sediments occur (Jankowski et al., 2017; Byrnes et al., 2019). This explains the observed trend of increasing RSLR toward the south as the thickness of Holocene sediment increases in that direction (fig. 2.16). However, subsidence resulting from sediment compaction has occurred throughout the delta's history. So, what caused the reversal from the delta's growth during the late Holocene to rapid inundation?

During floods, the Mississippi River channel and its abandoned distributaries overflow with sediment-laden water, spilling over levees and distributing sediments across the vast delta plain. Modification of this natural sediment dispersal system began over three centuries ago when a levee was constructed to prevent flooding in New Orleans. Since then, the levee system has been greatly expanded. In addition to flood control levees, 29 locks and dams have been built within the drainage network of the river, and an additional 8 on the Illinois River. These features have been tested on several occasions and have proven effective. Hurricane Katrina was a notable exception.

There is now a strong consensus within the scientific community that the reversal from the late Holocene growth of the Mississippi River Delta to its ongoing inundation is the result of accelerated sea-level rise coupled with alteration of the river's sediment delivery and dispersal system. According to the 2017 Coastal Master Plan for Louisiana, 11 billion to 34 billion tons of sediment would be needed to offset past and future land loss. Given current and predicted RSLR and the enormous quantities of sediment required to offset land loss, the scientific consensus is that saving the lower portion of the Mississippi Delta is no longer feasible.

Demise of the Brazos and Colorado Deltas

By the late Holocene, when the Mississippi Delta was constructing its six vast delta lobes on the Louisiana continental shelf, the Brazos, Colorado, and Rio Grande had declined in sediment supply and shifted from fluvial-dominated to wave-dominated deltas. This change occurred as the rate of sea-level rise decreased, but more arid conditions were occurring in west Texas, highlighting the role of climate in delta evolution. Erosion of the Brazos and Colorado deltas resulted in a prominent delta headland that divides the east Texas and central Texas coast (fig. 4.5). During historical times, the Brazos River has delivered enough sediment to the coast to maintain a wave-dominated delta, while the Colorado River nourishes a bayhead delta. The Brazos-Colorado delta headland is bounded to the east by Follets Island and to the west by the Matagorda Peninsula. Shoreline erosion rates are high on both barriers, indicating that these rivers are no longer significant suppliers of sand to the coast.

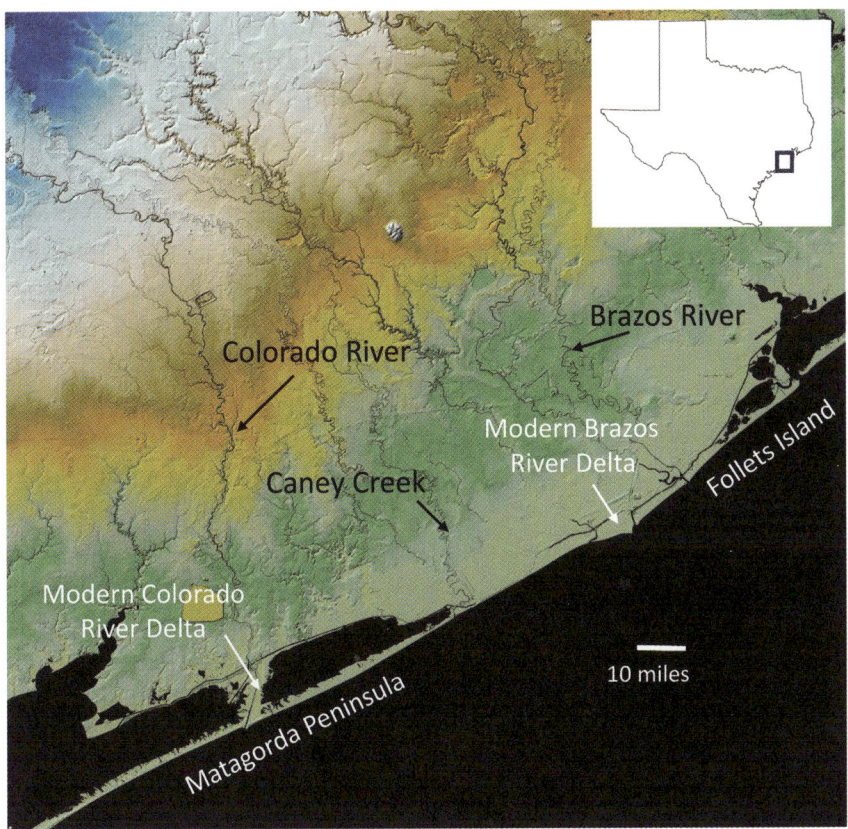

FIGURE 4.5. The Brazos-Colorado delta headland stretches nearly 50 miles and is a reminder that these combined rivers once played a crucial role in sediment delivery to the central Texas coast. Today, the Brazos River nourishes a relatively small wave-dominated delta, and the Colorado River has a bayhead delta. Erosion of the headland during the late Holocene provided sand for the construction of Follets Island to the east and the Matagorda Peninsula to the west. Today, both are rapidly eroding. Modified from Anderson et al. (2022).

The Brazos River nourished two wave-dominated deltas in historical times. Before 1929, the river entered the Gulf at Surfside Beach, which at that time was the site of the town of Velasco (fig. 4.6). Velasco was destroyed by a hurricane in 1884, killing most of its inhabitants. The town was re-occupied and its port at the mouth of the Brazos River reopened, but the expansion of the Galveston ship channel and its port facilities ultimately contributed to a decline in Velasco's economy and population. To revitalize the area's economy, in 1929 the river was diverted six miles to the west, and the former river mouth was converted to a ship channel and port facility (fig. 4.6). Over the next several decades the city of Freeport changed from a sleepy fishing community to one of the nation's largest coastal industrial complexes. So, the plan was an economic success but created one of the Gulf Coast's most vulnerable locations for hurricane surge and sea-level rise.

After the Brazos River was diverted, sediment supply to the pre-1929 delta ceased and it eroded rapidly, culminating in its destruction within

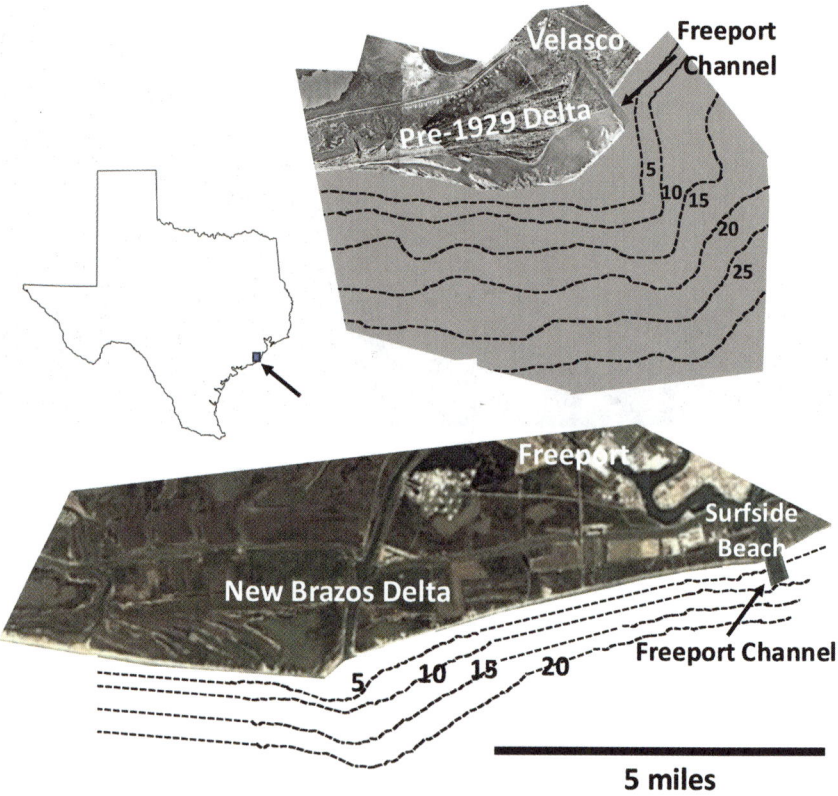

FIGURE 4.6. Historical development of the Brazos River delta. Water depths are in feet. The FIGURE at the top is an early 1930s aerial photograph showing the Brazos delta near the old town of Velasco. The FIGURE at the bottom is a 2010 Google Earth image showing the new Brazos delta. Note that the pre-1929 delta has been completely eroded. Modified from Rodriguez et al. (2000).

three decades. Sand that eroded from the delta was transported west by prevailing longshore currents to nourish the post-1929 delta. In only half a century, the new delta was roughly the size of the pre-1929 delta. This is an excellent example of the speed and efficiency of offshore wave erosion, which completely removed the pre-1929 delta and longshore sediment transport that nourished the new delta. It is also a good example of the rate and magnitude of human impact on the coast.

Twelve dams and reservoirs block sand delivery from most of the Brazos River's larger tributaries, but these dams are far enough upstream to have had little impact on short-term (decadal) sediment delivery to the delta. Most of the sand delivered to the delta comes from bank erosion and recycling of old channel deposits. The river diversion at Freeport, Texas, eliminated a six-mile stretch of the river from this process by rerouting the channel across muddy marshland. A detailed study of the Brazos delta in the late 1990s showed that since 1960, the delta had a decreasing growth rate, except for brief periods of punctuated growth following major floods (Rodriguez et al., 2000). In recent decades, most of the sand delivered to the delta has been deposited in sand ridges on the delta plain. That volume has decreased in recent decades as sand derived from erosion of the pre-1929 delta has decreased. Thus, while drier conditions have contributed to the long-term decline of the delta, its more recent decline was driven primarily by human influence. As for the longer-term prognosis, more arid conditions in the coming decades are expected to further diminish the river's sediment discharge, so the delta will continue its struggle for survival.

Before ~1 ka, the mouth of the Colorado River was approximately 25 miles east of its current location at Caney Creek (fig. 4.5). The river then experienced avulsion, abandoning the Caney Creek channel to occupy its current channel. A navigation chart constructed in 1839 shows the river flowing into east Matagorda Bay near its current location. There was no bayhead delta at this location (fig. 4.7). The Caney Creek delta had been eroded by this time, similar to the pre-1929 Brazos delta. In the late 1920s, a logjam was removed from the Colorado River, resulting in the rapid growth of a bayhead delta. By 1935, it extended across east Matagorda Bay. This was followed by the construction of a navigation channel through the Matagorda Peninsula to the Gulf of Mexico. However, the sediment supply from the river was too small to construct a wave-dominated delta. The establishment of the Lower Colorado River Authority in 1936 paved the way for the creation of seven dams and reservoirs over several decades. These combined alterations to the river's sediment delivery system have resulted in minimal delta growth and virtually eliminated it as a source of sand to the Texas coast.

a
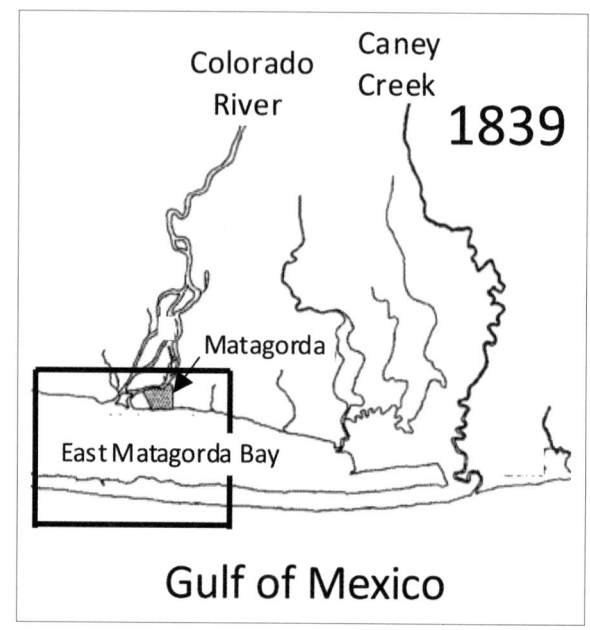

b
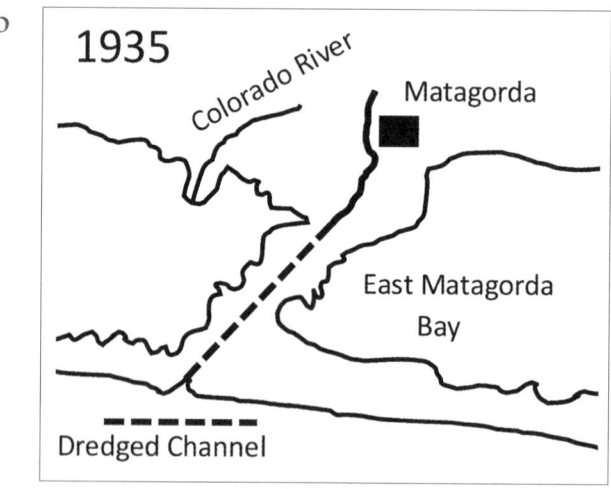

FIGURE 4.7. *Above and facing*, Historical charts showing the formation of the modern Colorado delta, Texas. Background map for 4.7 (c) from Google Earth.

c

The Demise of the Rio Grande Delta

Throughout the late Holocene, the Rio Grande maintained a wave-dominated delta (Banfield and Anderson, 2004). Its prominent delta headland is a reminder that it remained a formidable feature until recently (fig. 4.8a). This is consistent with the river's large drainage basin area, which is second only to the Mississippi River for rivers flowing into the Gulf of Mexico (fig. 3.3). However, increasing aridity in the Rio Grande's drainage basin during the late Holocene resulted in decreased sediment supply to the delta. The offshore delta was buried beneath the Texas Mud Blanket as it shifted landward.

A long historical record of precipitation in the southwestern United States, constructed from tree ring records, shows a reasonable correlation to the flow of the Rio Grande (Woodhouse et al., 2012) (fig. 4.9). Three extreme droughts (megadroughts) occurred around 1772–1776, 1892–1894, and 1953–1956 and correlate with La Niña events. Eugene Wahl and colleagues published a more recent record of precipitation revealing that the

a

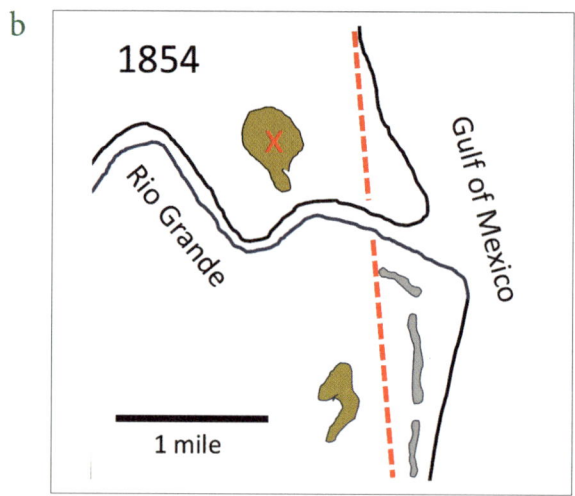

b

1854

Rio Grande

Gulf of Mexico

X

1 mile

FIGURE 4.8. *Above and facing,* (a) Google Earth image of the Rio Grande delta headland. The white box is the area shown in images (b) and (c). (b) 1854 navigation chart of the mouth of the Rio Grande showing a wave-dominated delta. (c) 2020 Google Earth image of the same area showing that landward migration of the shoreline has removed the delta and is now eroding the river channel. The red dashed line marks the same location in images (b) and (c).

c

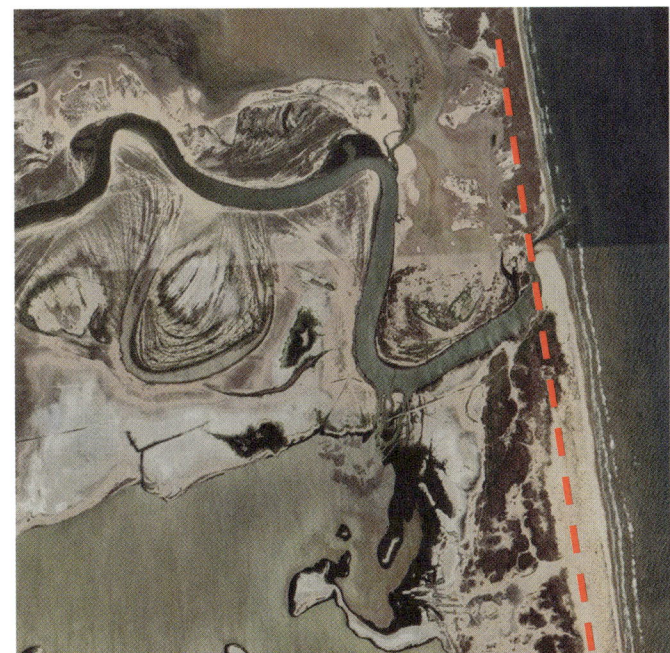

most recent drought conditions in the region began around 1990 and are a response to a historically unprecedented increase in temperature (Wahl et al., 2022). Indeed, 2000–2021 was the driest 22-year period of drought for at least 800 years (Williams et al., 2021). These drought conditions are expected to continue into the future, amplified by increasing temperatures and soil moisture depletion.

The long historical and paleoclimate records discussed above indicate that the early demise of the modern Rio Grande delta resulted primarily from increased aridity in the southwestern United States, which led to decreased river discharge. An 1854 navigation chart shows that the Rio Grande still nourished a wave-dominated delta with an asymmetric shape indicating dominant longshore sand transport toward the south (fig. 4.8b). Today, the delta no longer exists, and the river struggles to maintain access to the Gulf. The landward-migrating shoreline consumes remnants of the river's channel belt (fig. 4.8c). Shorelines on either side of the river mouth erode at rates ranging from 3 to 6 feet/year, estimated to be roughly six times the rate before 1850.

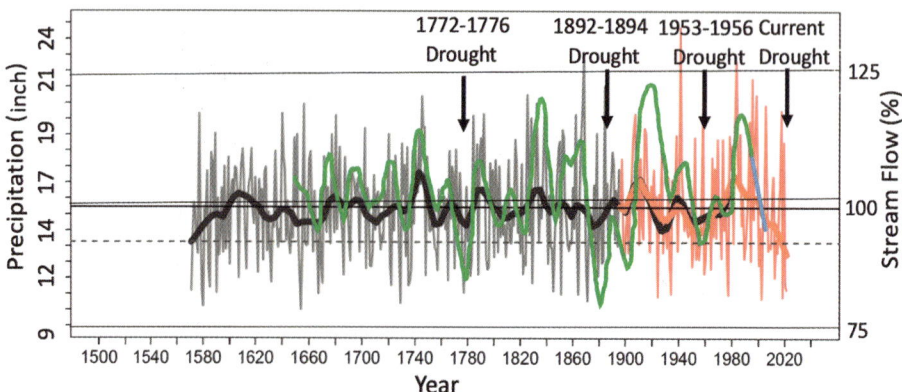

FIGURE 4.9. Combined historical records of precipitation (black and red lines) in the southwestern United States (modified from Wahl et al., 2022) and flow of the Rio Grande (green line) reconstructed from tree-ring data (modified from Woodhouse et al., 2012). Extreme droughts of 1772–1776, 1892–1894, and 1953–1956 are indicated. Note that the precipitation record extends to 2020 and shows the current drought, while the tree-ring record does not include the past decade.

The final demise of the delta was exacerbated by large-scale water usage for municipal and agricultural purposes, with a shift from less water-intensive crops like corn and cotton to more water-intensive fruits and vegetables following the 1994 passage of the North American Free Trade Agreement. In addition, water and sediment discharge has been altered by the construction of the Falcon Dam in the lower river basin (Benke and Cushing, 2005), and by the Marte R. Gómez and El Cuchillo Dams, which block flow from tributaries entering the river from the Mexican side of the border and south of the Falcon Dam. Today, the river's discharge is so low that it barely maintains access to the Gulf of Mexico.

Bayhead Deltas

Bayhead deltas are a vital component of the Gulf Coast, providing natural water filtration, an essential nutrient source, habitats and nursing grounds for a wide range of plant and animal life, and winter nesting grounds for waterfowl. They are one of the most vulnerable coastal settings to sea-level rise and climate change because they include expansive, low-elevation delta plains composed mainly of wetlands whose survival depends largely on the rate of RSLR versus the rate of sediment accumulation (vertical accretion). The wetlands drown if the vertical accretion rate does not keep pace with RSLR.

Studies of several Texas bayhead deltas showed that they experienced significant inundation during the early Holocene when the rate of sea-level rise exceeded 16 inches/century (Anderson et al., 2022). These studies also showed considerable inundation of bayhead deltas during the middle and late Holocene, when rates of sea-level rise were relatively low (average 2 to 6 inches/yr). These later flooding events are attributed to climate-induced reductions in sediment supply to these deltas. The current rate of sea-level rise is fast approaching early Holocene rates, and climate change threatens to reduce sediment to bayhead deltas in central and west Texas. Added to their propensity for rapid change is that many bayhead deltas have experienced direct human impacts on their freshwater inflows. This is especially true for some larger estuaries adjacent to metropolitan and industrial areas (fig. 4.10). The past and current responses of estuaries to sea-level rise and sediment supply are discussed further in chapter 6.

a

FIGURE 4.10. Gulf Coast bayhead deltas vary widely in size and shape. Images from Google Earth.

b

Escambia Delta

FIGURE 4.10 *continued*

c

Mobile Delta

d

FIGURE 4.10 *continued*

There has been a history of delta demise in the Gulf Coast region throughout the Holocene and into modern times. During the Holocene, the Brazos, Colorado, and Rio Grande deltas shifted from river-dominated to wave-dominated deltas in response to declining river sediment discharge, coinciding with the onset of more arid conditions in west Texas. The Brazos River has maintained a wave-dominated delta despite altering its course to a different location. Meanwhile, the Colorado River delta has been reduced to a bayhead delta. The Rio Grande has experienced alternating decadal-scale precipitation cycles since at least 1500. However, it maintained a wave-dominated delta until the twentieth century, when drought conditions, excessive water usage, and dam construction led to its final demise. The Brazos delta could suffer a similar fate in the coming decades. The Mississippi River Delta is the last of the great Gulf Coast deltas to experience demise caused mainly by human alteration of the river's sediment delivery and dispersal system, coupled with accelerating sea-level rise. Unfortunately, efforts to reroute sediment from the river back to wetlands are coming too late to save most of the delta. Several bayhead deltas will suffer significant inundation over the next few decades. These bayhead deltas are vital ecosystem components of Gulf Coast estuaries, and their demise will have a considerable impact.

During my many trips across Louisiana on Interstate 10, I looked forward to the drive between Lafayette and Baton Rouge, traveling across miles of the Mississippi Delta plain with its lush cypress swamps. As a geologist, I can't help but think about how virtually the entire coast from Mississippi to east Texas was once occupied by such a landscape. I have colleagues who spent their careers studying how deltas evolved over thousands of years. None of us realized that we would witness their demise.

5

Wave-Dominated Coastal Environments

Virtually the entire Gulf Coast is eroding, in some cases at alarming and unsustainable rates. Geological records indicate that this is a reversal in the long-term trend, as barrier islands, peninsulas, and chenier plains have had a history of stability and growth until recent times. Understanding the causes of this change and how to best prepare for it requires a basic understanding of coastal processes and how coasts responded to rising sea levels and climate change in the past.

Most of the Gulf Coast is lined by barrier islands, peninsulas, chenier plains, and mainland beaches, which collectively fall under wave-dominated coasts. In Texas, Louisiana, and Mississippi, large stretches of wave-dominated, sandy shoreline are inaccessible or have restricted access. Much of the Alabama and Florida coasts have been developed to the point that it is almost impossible to acquire unbuilt property with a view of the beach, or for that matter to drive along coastal highways where one can see the beach. In addition to being important urban and recreational areas, these wave-dominated coastlines are unique habitats, and they are the main line of defense for coastal bays and wetlands against rising sea levels and storm impact. In this chapter, we will discuss the processes that influence wave-dominated coasts, how and when these coastal environments evolved, ongoing changes that indicate a recent shift from growth to erosion and landward retreat, and the reasons for this change in coastal stability.

A coastal barrier is an elongated accumulation of sediment that parallels the coast and is at least partially separated from it by a lagoon or an estuary. This includes barrier islands and peninsulas, which together occupy over half of the coastline (fig. 1.3). The ability of these barriers to combat storm surge and sea-level rise increases with their width and height, which is largely a product of their age and manner of formation. Peninsulas are formed mainly through a process known as spit accretion, which begins with the formation of a sand spit at the mouth of an estuary, followed by elongation as longshore currents continue to nourish the downflow end of the spit (fig. 5.1a). Most modern barrier islands are believed to have been formed by landward migration during sea-level rise (fig 5.1b). They also form by breaching peninsulas.

The west Louisiana Chenier Plain is a unique wave-dominated coastal setting. It consists of 3-to-12-foot-high sandy ridges (*chenier ridges*) separated by marshes (fig. 5.2). They are composed of mud and sand from the Mississippi River that is exposed to erosion during severe storms to form cheniers. The eastern portion of the Chenier Plain has become inactive during the last few thousand years and is separated from the Gulf by wetlands and coastal lakes.

Mainland beaches make up a relatively small proportion of the Gulf Coast, where back-barrier bays and lagoons do not exist. They are primarily products of landward shoreline migration.

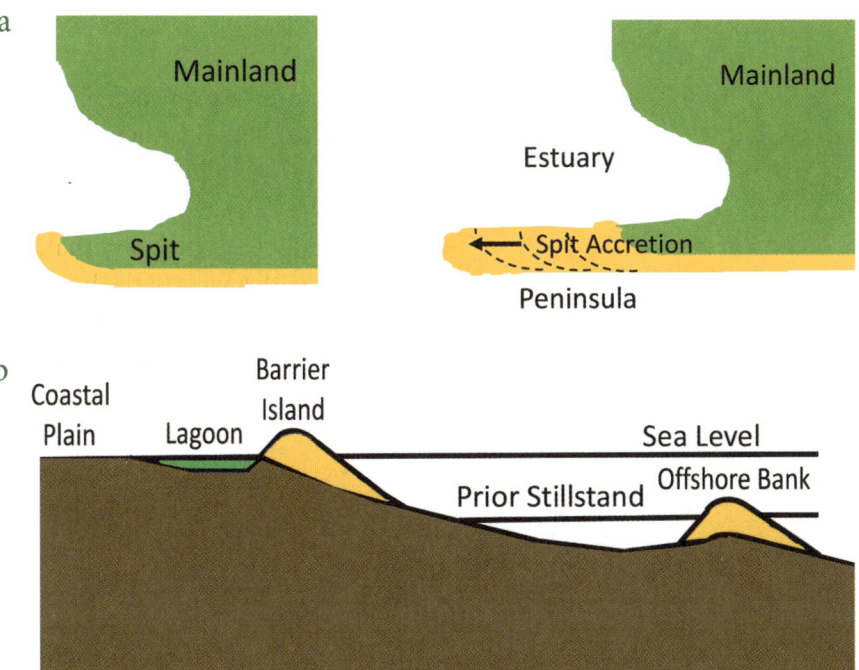

FIGURE 5.1. (a) Formation of a peninsula by spit accretion and (b) landward movement of coastal barriers during sea-level rise. The latter mechanism may partially preserve the older barrier as an offshore bank.

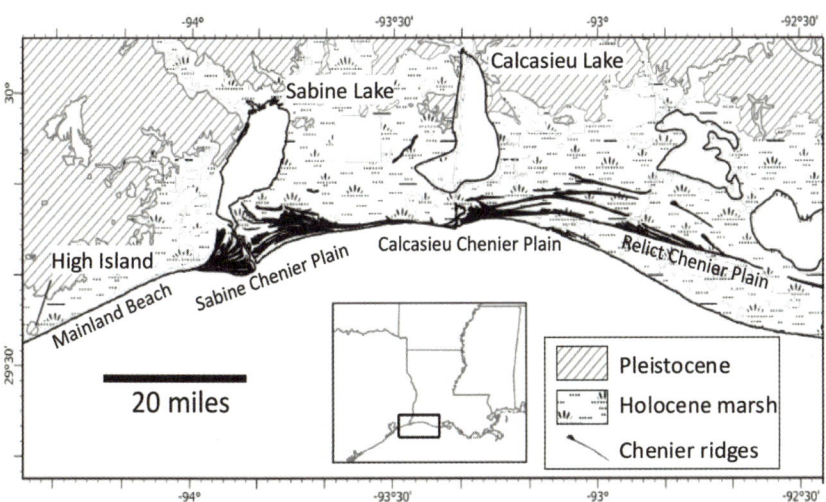

FIGURE 5.2. (a) The western Louisiana Chenier Plain is a unique coastal setting with alternating marshes and sand ridges. Modified from McBride et al. (2007).

FIGURE 5.2. (b) Google Earth image shows the Sabine chenier plain.

Some Basics: Wave-Influenced Coastal Processes

The Gulf Coast is a microtidal setting with less than two feet of tidal amplitude. As a result, wind-driven waves and currents dominate coastal processes. A basic grasp of these processes will help us understand how sediments move along the coast and how the coast responds to climate change and rising sea levels.

The magnitude of wave-driven coastal processes varies in response to day-to-day and seasonal weather patterns. Warmer months are generally characterized by persistent and mild onshore winds and modest wave conditions. During cooler months, the passage of strong meteorological fronts results in larger waves and more variable nearshore currents. These changing conditions strongly influence the movement of sand within the coastal setting.

In the open ocean, fair-weather waves (swell) are characterized by orbital water motion that diminishes with depth (fig. 5.3). In terms of wave geometry, it is the length of waves (distance between wave crests) that determines the depth at which they will first begin to drag bottom, which causes them to steepen and break. Along the Gulf Coast, fair-weather waves are typically less than 5 feet high and have 20-to-50-foot wavelengths.

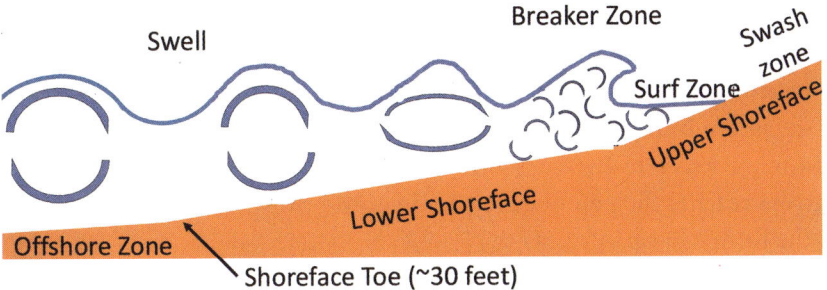

FIGURE. 5.3. Wave motion in waves approaching a coast. As waves move into shallower water, they experience bottom drag, and water particle motion within the wave takes on a more elliptical path. The velocity gradient of the landward-propagating wave increases to the point where surface water overtakes deeper water, and the wave breaks, resulting in a highly turbulent breaker zone and surf zone. Water movement on and off the beach occurs within the intertidal swash zone. In the Gulf Coast region, the depth at which storm waves begin to interact with the seafloor is commonly marked by a bathymetric change at depths between −15 and −30 feet.

The main factors regulating the wave regime of any coast are prevailing wind direction and speed and the offshore gradient, particularly the relatively steep portion of the seafloor near the coast where waves first begin to be influenced by bottom drag. This is the *shoreface zone*, and its gradient controls wave energy. The lower this gradient, the greater the energy dissipation as waves approach the coast, so a steep shoreface profile favors relatively high wave energy, and a gentle shoreface profile favors low wave energy.

As waves approach the shoreline and experience increased bottom drag, they become steeper as surface water flows faster than bottom water. Eventually, the surface water overruns the bottom water, and the wave breaks (fig. 5.3). This is the breaker zone, and its distance from the beach varies with changes in wavelength and height, shifting offshore as wave height and length increase. The breaker zone is typically within 300 feet of the shoreline in water less than 6 feet deep for fair-weather waves. Storm waves are generally 5 to 20 feet high and have 50-to-80-foot wavelengths, resulting in bottom drag in deeper water and a breaker zone farther offshore. Landward of the breaker zone is the surf zone, where breaking waves cause extreme turbulence. The most landward zone of wave influence is the swash zone, the intertidal zone where water swashes back and forth across the beach. The average width of the swash zone is controlled by the gradient of the beach and the tidal range.

Waves moving landward result in an onshore movement of water that produces flow along the coast in the form of a *longshore current* (fig. 5.4). Longshore currents play a crucial role in sand transport, a process known as *longshore transport*. As longshore currents flow along the coast, waves add more water to the current, causing it to accelerate. Eventually, the current reaches the point where energy must be dispersed by offshore flow in the form of a rip current.

Most of the Gulf Coast has relatively straight shorelines, but in some areas headlands extend offshore as shoals that are commonly associated with deltas, chenier plains, and ebb-tidal deltas. As waves approach the shallower shoals, they experience drag, which causes the waves to curve around them. This process, known as wave refraction, focuses wave energy on the headland (fig. 5.4).

The direction of the wind controls the direction of longshore currents. Since prevailing winds along most of the Gulf Coast are from the southeast, longshore currents flow predominantly from south to north along the west Florida coast, from east to west between the Florida Panhandle and east Texas, and from south to north along the southwest Texas coast (fig. 5.4b). Along the central Texas coast, there is a convergence of longshore currents flowing from the south and east. Over centuries to millennia, prevailing longshore current and longshore transport directions are recorded by lateral migration of coastal barriers and tidal inlets. Historical records of longshore transport are manifest by the accumulation of sand on either side of manufactured features and reveal flow and transport directions similar to the long-term record revealed by longshore spit and barrier accretion (fig 5.1a). This indicates that longshore transport patterns have stayed mostly the same during the last few thousand years.

Tidal Inlets and Tidal Deltas

Tidal inlets and *tidal deltas* are integral to wave-dominated coasts because they sequester sand that moves within the longshore transport system. They also play a crucial role in controlling the influx of saline Gulf of Mexico waters to estuaries and regulating estuarine circulation. Most large inlets have been modified by the dredging of navigation channels and by the construction of jetties. The result has been the disruption of sand transport, deposition, and alteration of estuarine salinities.

a

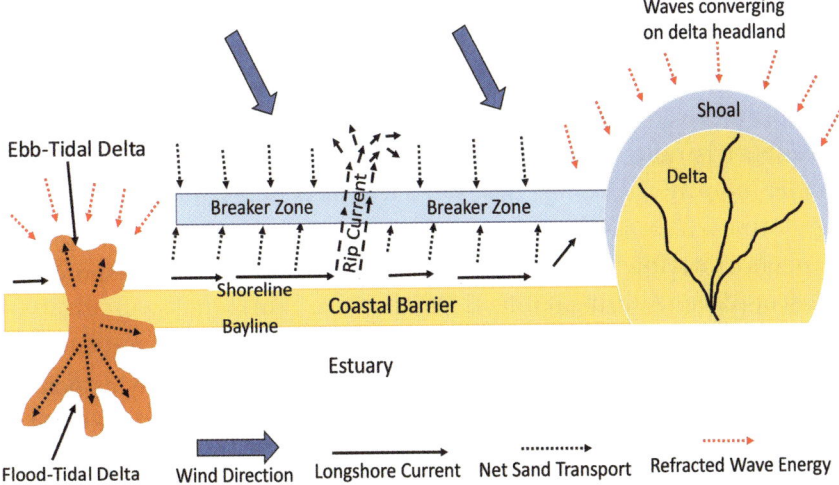

Waves converging on delta headland

Ebb-Tidal Delta

Shoal

Delta

Breaker Zone

Rip Current

Breaker Zone

Shoreline

Coastal Barrier

Bayline

Estuary

Flood-Tidal Delta — Wind Direction — Longshore Current — Net Sand Transport — Refracted Wave Energy

b

Longshore Transport

Rock Jetty

FIGURE 5.4. (a) Circulation in wave-dominated coastal environments. Wave-driven circulation is characterized by onshore flow seaward of the breaker zone and offshore and alongshore flow landward of the breaker zone. The alongshore flow (longshore current) accelerates until its energy is dissipated by turning offshore into deeper water as a rip current. Waves approaching a headland are refracted, focusing their energy on the headland. Tidal inlets separate coastal barriers and locations where longshore transport is disrupted by flow into and out of an estuary. Sand transported into the estuary during flood tides accumulates in a flood-tidal delta, and sand moved offshore during ebb tides is deposited in an ebb-tidal delta. (b) Aerial photograph of Galveston Island showing sand accumulation by a rock jetty that blocks southward longshore transport, resulting in the growth of beaches near the jetties and the erosion of beaches downstream.

Tidal inlets of the Gulf Coast generally fall into two categories: those connecting river-influenced estuaries with the Gulf and those associated with shore-parallel estuaries and lagoons. The latter are smaller and have shorter lifespans, especially those formed by breaching coastal barriers during storms. When sand moving within the longshore transport system encounters a tidal inlet, it is transported into the estuary by rising tides (flood tides) or offshore during falling tides (ebb tides). This sand accumulates in flood-tidal and ebb-tidal deltas, which can be a significant component of the coastal sand budget (fig. 5.4). Ebb-tidal deltas are exposed to offshore waves and currents and are therefore part of the longshore transport system. In contrast, flood-tidal deltas are protected from these forces and sequester sand, thus removing it from the longshore transport system.

The most extensive inlets and tidal deltas are associated with large river-influenced estuaries, such as Mobile Bay and Galveston Bay (fig. 5.5a, b). These long-lived inlets and tidal deltas grew significantly as adjacent barrier islands and peninsulas evolved, gradually robbing the adjacent coast of sand. The Bolivar tidal delta between the Bolivar Peninsula and Galveston

a
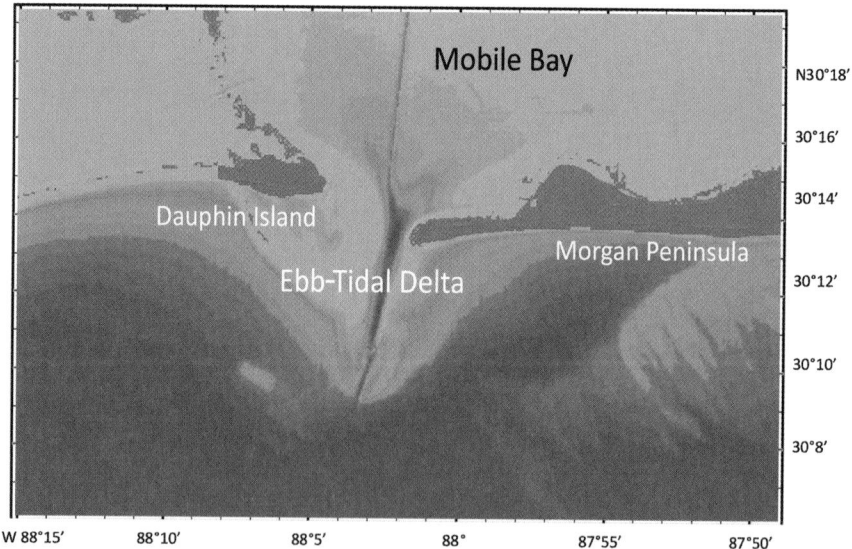

FIGURE 5.5. *Above and facing*, Tidal inlets of (a) Mobile Bay, (b) Galveston Bay, and (c) Sabine Lake are all characterized by large ebb-tidal deltas and relatively small flood-tidal deltas, in contrast to (d) the San Luis Pass tidal delta, which separates Galveston Island from Follets Island and is dominated by its flood-tidal delta. Images (a)–(c) from GeoMapApp; image (d) from Google Earth.

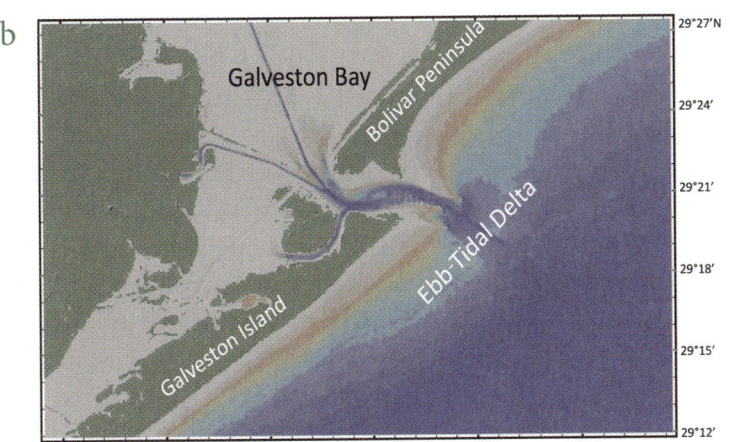

b

Galveston Bay

Bolivar Peninsula

Ebb-Tidal Delta

Galveston Island

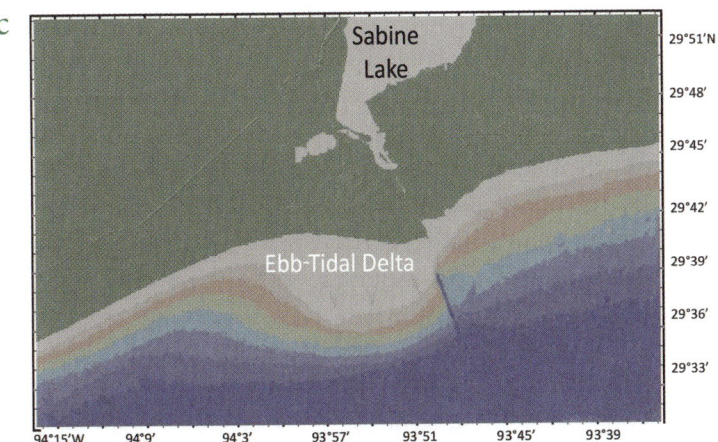

c

Sabine
Lake

Ebb-Tidal Delta

d

Flood-Tidal Delta

Galveston Island

Follets Island

Island is estimated to contain most of the sand that would have nourished Galveston Island in the past 2,500 years.

Most of the larger inlets on the Gulf Coast have been altered by the construction of navigation channels and associated rock jetties that have disrupted natural sediment transport. Sand transport along the Galveston Island shoreline changed in the early 1900s when the Houston/Galveston Ship Channel was dredged, and rock jetties were constructed on both sides of the channel (fig. 5.6). These jetties trapped most of the sand moving from east to west along these barriers. Smaller jetties were also constructed along the Galveston Seawall. These jetties block what little sand remains in Galveston's longshore transport system (fig. 5.4b).

Is There a Gulf Coast Sand Deficit?

During the more humid conditions of the late Pleistocene and early Holocene, rivers delivered enough sediment to the Gulf Coast to form large deltas on the continental shelf from west Florida to south Texas (fig. 1.5). Higher rates of subsidence on the outer shelf promoted preservation of these deltas (fig. 2.13). As sea level continued to rise, the deltas shifted onto the inner shelf where subsidence rates were lower, promoting their erosion and producing most of the sand of the modern coast. Offshore of west Florida and Alabama, the Apalachicola, Perdido-Escambia, and Alabama-Tombigbee deltas were eroded by waves and spread across the inner continental shelf to form the MAFLA Sheet Sand (fig. 1.7), which became the dominant source of sand for the northwest Florida and Alabama coast. To the west, the vast St. Bernard lobe of the Mississippi Delta occupied the continental shelf until ~2.0 ka (fig. 4.1). Subsequent delta erosion produced much of the sand that makes up the modern Mississippi barrier island chain (Otvos, 2018). Additional sand was derived from erosion of an extensive offshore channel network (fig. 1.6b). The Teche and Lafourche delta lobes were eroded to produce sand for the Louisiana barrier islands. Mud eroded from these deltas was transported west, where alternations in mud and sand supply and coastal erosion during storms resulted in chenier plain development. In Texas, a shift to more arid conditions during the late Holocene resulted in decreased sediment supply from rivers and abandonment of offshore deltas. Subsequent erosion of these deltas yielded most of the sand for the Texas coast. This offshore sand supply decreased as shoreline migration

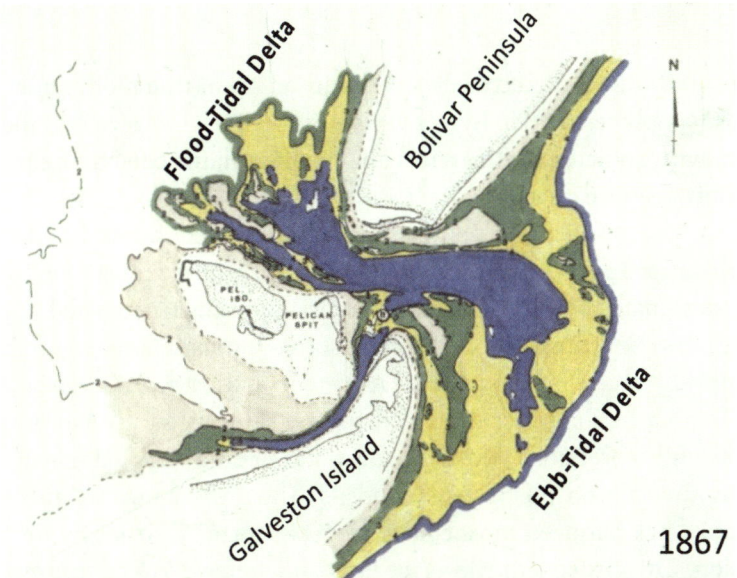

a

Flood-Tidal Delta

Bolivar Peninsula

N

PEL
180.

PELICAN
SPIT

Galveston Island

Ebb-Tidal Delta

1867

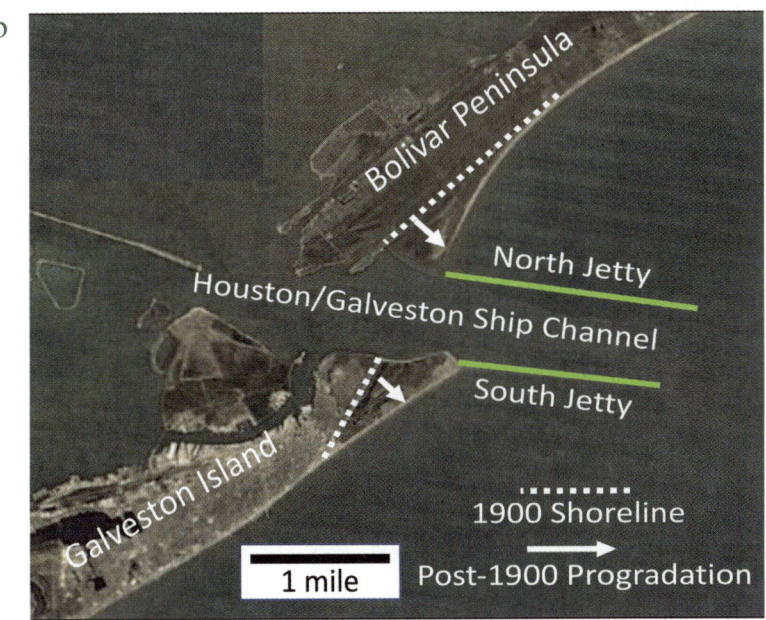

b

Bolivar Peninsula

North Jetty

Houston/Galveston Ship Channel

South Jetty

Galveston Island

1900 Shoreline

Post-1900 Progradation

1 mile

FIGURE 5.6. Construction of the Houston/Galveston Ship Channel and associated rock jetties in the early 1900s altered the movement of sand in the longshore transport system and reduced sand delivery to the tidal delta. As a result, the large ebb-tidal delta that existed in 1867 (a) was eroded, and much of that sand was deposited on the north and south sides of the jetties, where beaches have advanced offshore since jetty construction. Background image (b) from Google Earth.

slowed and offshore sand bodies were buried in marine mud. Differences in sand supply resulted in highly variable shoreline change across the Gulf Coast, with some coastal barriers and chenier plains continuing to grow while others were eroding.

The processes responsible for the erosion of offshore sand bodies and the onshore transport of sand during a transgression occur within the wave-dominated shoreface. Shoreface erosion is well documented in Texas, where multiple sediment core transects across the shoreface and inner continental shelf have been acquired (Anderson et al., 2022). Cores collected at water depths shallower than 24 to 30 feet sampled sandy shoreface deposits, while cores collected at greater water depths sampled marine mud resting directly on Pleistocene deposits. Thus, the landward-migrating shoreface has removed most coastal deposits formed earlier in the transgression. This process of shoreface erosion is referred to as transgressive ravinement because it is widespread and highly efficient, much like a giant bulldozer plowing the continental shelf as the shoreline migrates landward.

Most of the sand produced by landward migration of the transgressive ravinement surface is recycled back into the coast. However, a decrease in the rate of coastal migration during the late Holocene resulted in less sand being delivered to the coast from offshore and more sand being sequestered in tidal deltas and in *washover* deposits, which are composed of sand that washes across the barrier during hurricanes (fig. 5.7). If the barrier is breached during storms, sand is concentrated in *washover fans*. Sand is also transported offshore during hurricanes, but most sand is transported back to the coast following a storm. The exception is sand transported seaward of the shoreface, but studies of offshore Texas have found this to be mostly very fine sand eroded from the lower shoreface (Odezulu et al., 2020). Humans have contributed to the loss of sand from the coast by altering the river sand supply and disrupting the longshore dispersal system with hard structures.

Studies in Texas have shown that sand sequestration in flood-tidal deltas increased during the late Holocene as coastal barriers and the tidal inlets separating these barriers evolved. A detailed investigation of the San Luis Pass tidal delta (fig. 5.5d) showed that most of the sand eroded from Galveston Island during the past ~2,000 years is accounted for by the growth of the flood-tidal delta (Wallace and Anderson, 2013).

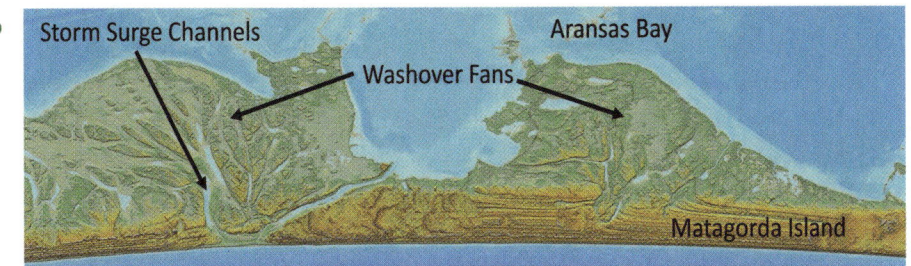

FIGURE 5.7. (a) Image of Matagorda Peninsula, Texas, showing washover deposits behind this narrow coastal barrier. (b) Image showing two large washover fans near the west end of Matagorda Island, Texas. Note storm surge channels where the island was breached. LiDAR data from NOAA.

The current trend across the Gulf Coast is diminishing sand supply while sea-level rise is accelerating. Given this observation, there is a dire need for sand budget analyses to document these changes and determine the volumes of sand needed to balance *coastal sand budgets*. One approach in sand budget analysis involves numerical modeling. However, it relies on assumptions about sand supply, dispersal, and deposition that commonly lead to unrealistic results. An alternative is the geological approach, which relies on radiocarbon ages from sediment cores to measure long-term sediment transport and accumulation rates. These longer-term records are needed to assess the impact of climate change and sea-level rise on sand supply and dispersal. Only Louisiana has a coastal management program that has rigorously acquired sand budget information. Mississippi, Alabama, and Florida need beach nourishment, but these states are fortunate to have significant offshore sand resources. Texas is less fortunate because offshore sand bodies are primarily buried in mud. Despite this, the Texas General Land Office, the agency responsible for coastal preservation in the state, is moving forward with plans to combat coastal erosion without knowing how much sand is needed and where ample sand resources exist (see chapter 9).

Understanding and Predicting Coastal Change Using Numerical Models

An important objective of coastal scientists is to better understand the factors that regulate coastal change, how they have varied in the past, and how they are likely to vary in the future. Early efforts to predict coastal response to rising sea levels commonly relied on simple "bathtub" models where coastal topography was flooded using different magnitudes of sea-level rise. However, the coastal science community has long recognized that coastal response to rising sea levels is influenced by multiple factors, including sand supply, wave climate, barrier width and height, shoreface profile, offshore and onshore sand flux, overwash sand flux, and sand thickness within the barrier and shoreface (fig. 5.8). As we will see, shoreline migration on Gulf Coast barriers was highly variable during the late Holocene, even in areas where rates of sea-level rise were similar across a region. This variability is consistent with numerical models that reveal

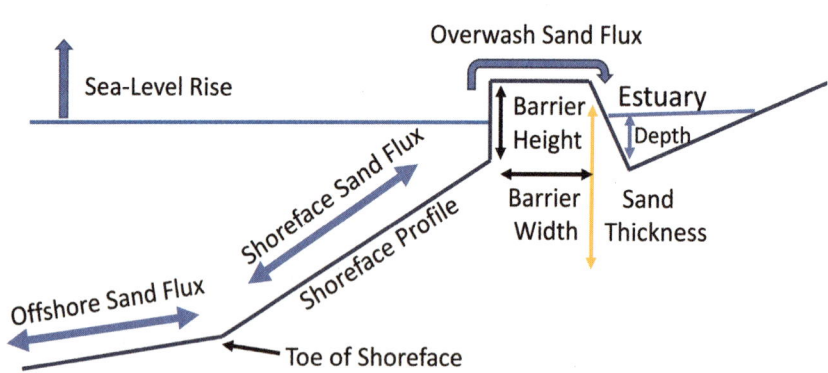

FIGURE 5.8. This FIGURE illustrates the range of variables used in numerical models to predict coastal barrier response to sea-level rise.

a complex and time-variable response of coastal barriers to sea-level rise (e.g., Lorenzo-Trueba and Ashton, 2014).

In general, shoreline stability increases with barrier width and height because wider and higher barriers are less susceptible to storm washover. Thicker barriers are more stable than thin barriers because more sand is recycled from the retreating shoreface back into the barrier. Quantifying these variables is the first step in forecasting shoreline change along any portion of the coast. The next step is establishing the sand budget, which calls for a reliable estimate of onshore and longshore transport rates, storm sand removal rates, and sand sequestration in tidal deltas. Unfortunately, sand budget analysis is time consuming and costly, which explains why this crucial information is lacking for most of the Gulf Coast.

Historical Records of Shoreline Change

To assess the response of coastlines to accelerating sea-level rise and other factors, it is helpful to relate historical shoreline changes to those during prehistoric times. Historical records of coastal change are based on early navigation charts and aerial photographs. Texas has been a leader in this field, with the publication of a series of reports by the Texas Bureau of Economic Geology in the 1970s describing historical shoreline changes along the coast. This program continues today and relies on more sophisticated measurements using various new techniques and data types, including LiDAR imaging, which uses laser ranging and GPS technology to measure elevation changes.

In 2004, the US Geological Survey (USGS) assumed the role of monitoring historical shoreline change along the Gulf Coast. Gulf-wide estimates of shoreline changes are based on "shore-normal transects with associated long-term rates of shoreline change" (available for download at https://pubs.usgs.gov/of/2004/1089/gis-data.html). Unfortunately, those data extend only through 2001. Since 2001, monitoring of shoreline migration has again shifted back to state agencies with different methods and levels of detail. Despite these differences, the results from state and federal programs have revealed high rates of shoreline migration across the Gulf Coast. Unfortunately, these historical records do not extend far enough back in time to fully assess the causes of coastal change and how the coast has responded to faster rates of sea-level rise and climate change.

To accomplish this, it is necessary to study coastal change over centennial to millennial timescales using geological methods.

Geological Records of Coastal Change

Shorelines tend to migrate seaward when sand supply outpaces the rate of sea-level rise, a process known as *progradation.* If sand supply keeps pace with sea-level rise, the shoreline remains fixed for long periods and grows vertically, a process known as *aggradation. Transgression* occurs when sand supply does not keep pace with sea-level rise and the shoreline migrates landward. Mississippi barrier islands represent a fourth category of coastal barrier evolution where significant lateral (shore-parallel) migration occurs. These are referred to as laterally accreted barriers, and the current rates of change for these barriers are believed to be unsustainable at timescales of decades to a few centuries. A key objective of coastal geologists is to assess the past and current stages of development for different coastal barriers and study how they responded to changes in the rate of sea-level rise and sand supply. This is done using geomorphological features and sediment cores.

Geomorphological Records of Coastal Change

Seaward and lateral growth of shorelines are commonly marked by linear *beach ridges*, which are easily recognized in aerial photographs and LiDAR images (fig. 5.9). Beach ridges form and gain elevation when sand is transported onshore during and following storms and when wind-blown sand is captured. As the shoreline migrates seaward, new ridges are formed, adding width and elevation to the barrier. The result is sets of beach ridges that decrease in age in an offshore direction, providing a means of measuring rates of barrier growth. Some Gulf Coast barriers display multiple beach ridge sets with different orientations, recording pauses and directional changes in barrier growth (fig. 5.9). Lateral (parallel to the coast)

FIGURE 5.9. *Facing,* Examples of beach ridges. (a) Oblique aerial view of St. Vincent Island, Florida, showing beach ridges that record seaward growth of the island. Low areas between ridges (swells) contain freshwater marshes. From Google Earth. (b) LiDAR image of a portion of Galveston Island, Texas, showing two beach ridge sets with different orientations that record two stages of barrier growth. Storm washover features record an early stage of barrier development when the island was narrower. From NOAA. (c) Sanibel Island, Florida, is characterized by beach ridges that curve to the north and record lateral accretion through inlet migration. From Google Earth.

a

b

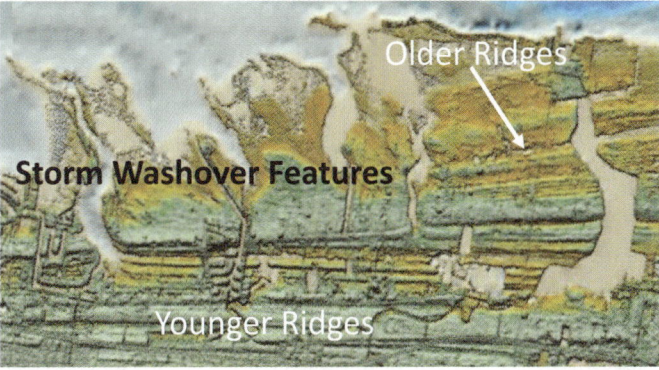

c

growth of barrier islands and peninsulas is marked by beach ridges that curve landward in the direction of inlet migration (fig. 5.9c). Episodes of transgression are marked by truncated beach ridges, storm surge channels, and washover fans (fig. 5.7).

Several Gulf Coast barrier islands and peninsulas have beach ridges overprinted by washover channels and fans, indicating that the barrier went through a phase of progradation followed by transgression (fig. 5.7). Other coastal barriers have washover features on their landward side and beach ridges on their seaward side, indicating an early transgressive history followed by progradation (fig. 5.9b).

Stratigraphic Records of Shoreline Migration

Aerial photographs like the ones in figure 5.9 can be used to guide the acquisition of sediment cores needed to determine the age of individual ridges and establish the timing and rate of barrier growth. Likewise, cores from storm washover features help constrain timing and transgression rates. Offshore cores provide additional information about shoreface advance and retreat timing and magnitude. This stratigraphic analysis is possible because coastal barriers and mainland shorelines are composed of discrete onshore and offshore environments with different physical processes that result in unique sedimentary deposits known as *sedimentary facies*. These sedimentary facies are distinguished by their grain size, sedimentary structures, and fauna. The stratigraphic stacking patterns of sedimentary facies are used to determine shoreline progradation, aggradation, and transgression episodes.

Figure 5.10 illustrates different coastal zones and associated sedimentary facies. The subaerial barrier comprises mainly beach ridges and eolian dunes. The *bayline* is where a coastal barrier meets the estuary or lagoon, and washover deposition occurs. The storm beach extends from the fair-weather swash zone to the storm swash zone, associated with the youngest beach ridge. The swash zone is characterized by thin layers of sand (sand laminae) that slope gently offshore at the angle of the foreshore slope. It is further characterized by a diverse assemblage of mollusk shells, most of which come from offshore. One exception is *Donax*, a small mollusk that spends its life burrowing into the sand only to be exposed to wave swash. This lifestyle explains why *Donax* is sometimes called the "neurotic clam." Because *Donax* lives in the swash zone, it is used as a sea-level indicator when it occurs in laminated sand or matching shell pairs.

a

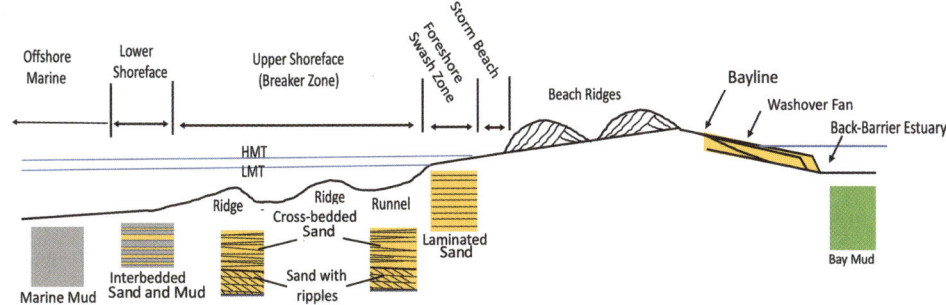

Offshore Marine

Lower Shoreface

Upper Shoreface (Breaker Zone)

Foreshore

Swash Zone

Storm Beach

Beach Ridges

Bayline

Washover Fan

Back-Barrier Estuary

HMT

LMT

Ridge

Ridge

Runnel

Cross-bedded Sand

Laminated Sand

Marine Mud

Interbedded Sand and Mud

Sand with ripples

Bay Mud

b

FIGURE 5.10. (a) Wave-dominated coastal environments and sediment transport zones. Colored boxes represent different sedimentary facies that distinguish these environments based on sedimentary structures and sediment grain size (gray = offshore mud, yellow = sand, and green = bay mud). (b) These coastal environments are also distinguished by their unique fauna, including *Donax*, a mollusk that inhabits the swash zone.

Seaward of the swash zone is the upper shoreface zone, where breaking waves and associated wave turbulence dominate. It is the area of active sand transport during fair-weather conditions and lies within a few hundred feet of the beach in water depths of less than 6 to 12 feet. Beneath the breaker zone, grains are swept from onshore and offshore to construct a sandbar (ridge). The ridge will migrate landward as wave energy decreases, while new ridges are constructed farther offshore when wave energy increases. This commonly results in more than one ridge at a given time and more than one breaker zone. Shore-parallel depressions between ridges are called runnels. A diverse assemblage of mollusk shells and shell debris also characterizes the shoreface. Over months and years, sand within the upper shoreface moves alongshore, transported by prevailing longshore currents, with landward and seaward oscillations in the ridge and runnel system resulting in interbedded ripples and landward-dipping cross-beds that record landward migration of ridges as waves diminish in size.

Seaward of the upper shoreface at water depths between ~10 and ~30 feet, bottom sediments are characterized by interbedded mud and sand with scattered shell material. Sedimentation in this zone records alternations between fair weather and storms. This is the lower shoreface zone (fig. 5.10). Seaward of the lower shoreface zone is the offshore marine zone. It is characterized by mud with a diverse assemblage of mollusk shells and thin sand layers that record storm events.

The unique sedimentary facies described above and illustrated in figure 5.10 allow sediment cores collected from the beach and offshore to be used to distinguish regressive and transgressive stratigraphic sequences and thus reconstruct shoreline migration through time. Regressive shorelines have a stratigraphic sequence indicative of shallowing upward depositional environments formed as the shoreline prograded, with marine and lower shoreface sediments overlain by upper shoreface and foreshore deposits (fig. 5.11). A deepening upward sequence of depositional environments characterizes transgressive shorelines. However, the beach and shallow water facies are typically eroded by storm waves as the shoreline migrates landward over decades to centuries. This is the *transgressive ravinement surface*, which usually manifests as marine mud resting directly on lower shoreface deposits or Pleistocene deposits. In cases where the transgression has occurred rapidly, shoreface deposits are thin and overlie back-barrier washover and estuarine deposits.

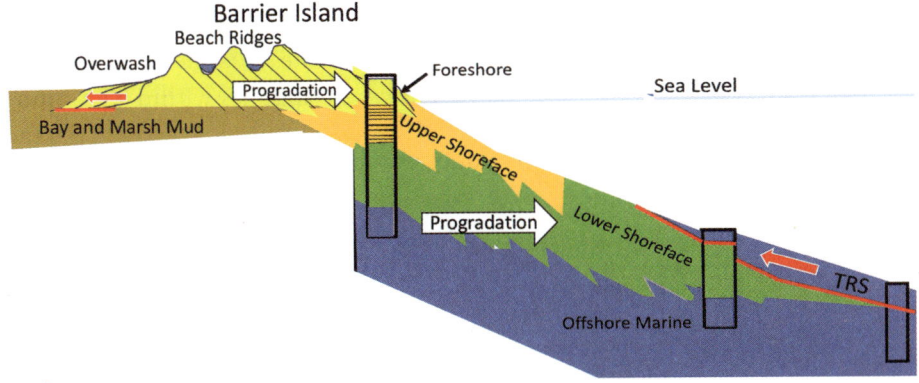

FIGURE 5.11. Idealized stratigraphic section that illustrates how sediment cores are used to reconstruct shoreline migration through time. The white arrows designate shoreline progradation, indicated by beach ridges and the foreshore and upper shore-face sedimentary facies resting on lower shoreface and marine sedimentary facies. The red arrows designate a recent episode of shoreface retreat (transgression) marked by a landward shift (onlap) of marine mud (TRS = transgressive ravinement surface, shown by the red line) and by storm overwash that occurs as the barrier becomes lower and narrower.

Generally, thicker coastal barriers are more stable because sand is recycled from the eroding barrier back into the coastal sand transport system as the shoreline migrates landward. Barrier thickness is determined by the relief of the surface upon which the barrier rests (*antecedent topography*) and by sand supply. The thickest Gulf Coast barriers fill old tidal inlets, commonly up to 50 feet thick. Coastal barriers become thinner during transgression because shoreface erosion yields less sand to nourish the retreating shoreline. Ultimately, most of the sand eroded from the barrier is deposited as washover. This stage of the barrier aging process is called *rollover* and marks the final stage of a barrier's existence. Some good examples of barriers that have reached this stage include Follets Island, the Matagorda Peninsula, and South Padre Island in Texas, and Dauphin Island in Alabama. These barriers are typically less than five feet thick.

Radiocarbon dating and *optically stimulated luminescence* (OSL) dating of coastal deposits provide age constraints that allow determination of the actual timing and rates of coastal change. Improvements in instrumentation and methodology over the past three decades have resulted in greater resolution for both methods. Latest-generation instruments allow radiocarbon dating of much smaller amounts of shell material with greater precision than was previously possible. The OSL method measures doses from ionizing radiation within sand grains that decay over time once the grains are buried and no longer exposed to light. Where applied, it has yielded results consistent with radiocarbon dating. So, it is now possible to measure barrier progradation and retrogradation rates with an accuracy of decades to a few centuries, determine sediment transport and deposition rates for constructing long-term sediment budgets, and compare these long-term rates to changes that occurred during historical time.

Coastal Barrier Evolution

Modern coastal barriers and chenier plains were formed mainly during the past 4,000 years when the rate of sea-level rise was less than ~2.0 inches/century (0.5 mm/yr). During this time, sand supply to the coast decreased as offshore sand sources were overrun by the landward-migrating shoreline and buried in mud. As a result, coastal barriers continued to change significantly as onshore sand supply became more crucial to sustaining the coast. More recently, humans have altered the supply of sand from rivers and its distribution along the coast. Meanwhile, the rate of sea-level rise in the Gulf of Mexico has accelerated, shifting the scales of coastal change in favor of shoreline retreat. In the following sections, we will discuss examples of how coastal barriers and chenier plains evolved during the late Holocene (~4 ka to 1850 CE), relying on the methods above. We will then compare these changes to those that have occurred since the mid-nineteenth century, using rates measured by various state agencies and the USGS. We begin with case studies from Texas, where the most detailed records of barrier evolution exist.

Texas Coastal Barriers

Barrier islands and peninsulas span 81 percent of the Texas coast (~310 miles), making it North America's longest stretch of coastal barriers. These include Padre Island, which, at 112 miles in length, is the world's longest barrier island, as well as Galveston Island, one of the world's most populated barrier islands. However, the vast majority of the coast remains relatively unaltered by direct human influence.

Texas is in many ways an ideal natural laboratory for research in coastal geology because of its variable climate and oceanographic setting and because the preservation of onshore and offshore Holocene deposits exceeds that in other areas of the Gulf Coast. It includes progradational, aggradational, and transgressive coastal barriers, providing an opportunity to study these different barrier types and how they have responded to sea-level rise, changes in sand supply, and other factors. Also, Texas has a history of coastal research dating back to the late 1950s, including several studies of coastal barriers and their evolution. The results reveal different styles and timing of barrier development and different responses of these barriers to sea-level rise and other factors (Anderson et al., 2022). In addition, Texas has the longest history of monitoring historical shoreline change, with records spanning nearly a century (fig. 5.12). These historical records

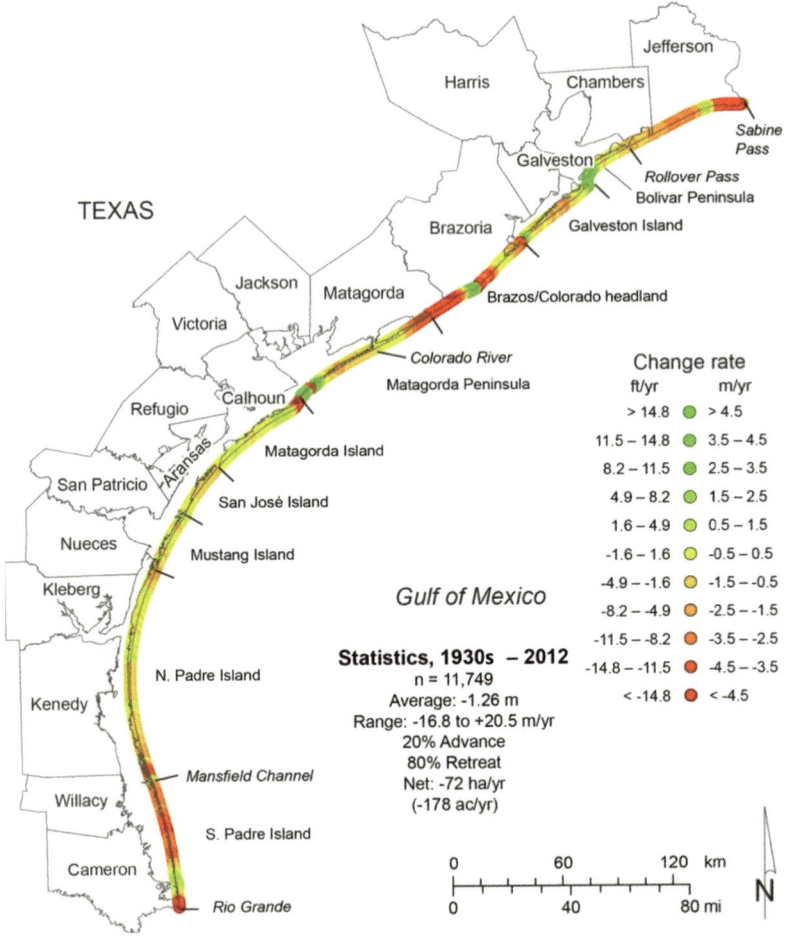

FIGURE 5.12. Rates of shoreline change in Texas were calculated from shoreline locations between the 1930s and 2012. From Paine et al. (2014). Reproduced with permission of the Gulf Coast Association of Geological Societies.

show that landward shoreline migration has occurred along 84 percent of the Texas coast, with an average rate between the 1930s and 2012 of − 4 feet/year (Paine et al., 2014). Combining these geological and historical records provides insight into how coastal barriers have responded to natural and anthropogenic influences and assesses their relative impacts on the coast.

Bolivar Peninsula

Research into the evolution of the Bolivar Peninsula was conducted over several years. It included the collection of seismic profiles, *ground-penetrating radar* (GPR) data, numerous sediment cores, and radiocarbon ages that have yielded a stratigraphic record of the peninsula's history. During the 1990s, there were several operating sand pits on the peninsula. While it was not a good idea to mine the peninsula for sand to nourish its beaches, these pits provided a rare opportunity to observe firsthand the anatomy of the barrier and calibrate these findings with those acquired using GPR images to establish a coring strategy for future work.

The Bolivar Peninsula began as a spit on the east side of Galveston Bay that accreted toward the west under westward-flowing longshore currents. The original peninsula was a narrow and low barrier that increased in thickness as it accreted toward the west, filling the deep Bolivar tidal inlet (fig. 5.13a). The bayside of the peninsula is dominated by washover fans that record an early phase of transgression when the barrier was low and narrow. Ultimately, the barrier began to prograde, gaining width and elevation through beach ridge development. Two sets of beach ridges exist. An older shore-parallel set of ridges extends along the central part of the peninsula and curves northward near its west end, where it fills the Bolivar tidal inlet. Radiocarbon ages from this ridge set show that this early stage of progradation occurred between ~2.8 ka and ~1.2 ka (Rodriguez et al., 2004).

A younger set of beach ridges trends parallel to the coast and curves seaward near the western end of the peninsula. This younger beach ridge set yielded radiocarbon ages younger than 0.65 ka. The gap in the age of these ridges indicates that the peninsula eroded after ~1.2 ka and before ~0.65 ka. The interpretation is that this hiatus was formed by one or more hurricanes that virtually decapitated the barrier, creating a prominent shell layer sampled in sediment cores and exposed in sand pits (fig. 5.13). This interpretation is further supported by evidence for a dramatic increase

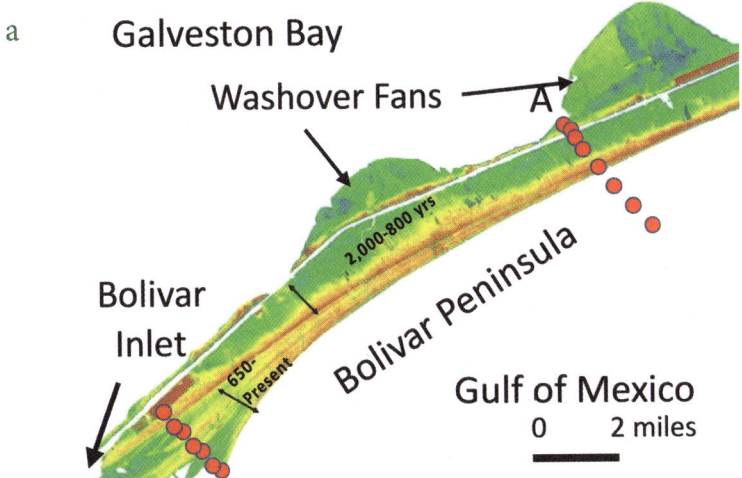

a

Galveston Bay

Washover Fans ➔ A

2,000-800 yrs

Bolivar
Inlet

650-Present

Bolivar Peninsula

Gulf of Mexico

0   2 miles

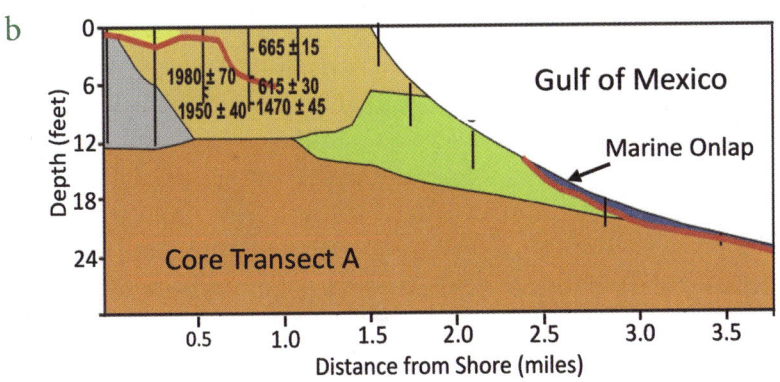

b

Depth (feet)

665 ± 15

1980 ± 70   615 ± 30
1950 ± 40   1470 ± 45

Gulf of Mexico

Marine Onlap

Core Transect A

Distance from Shore (miles)

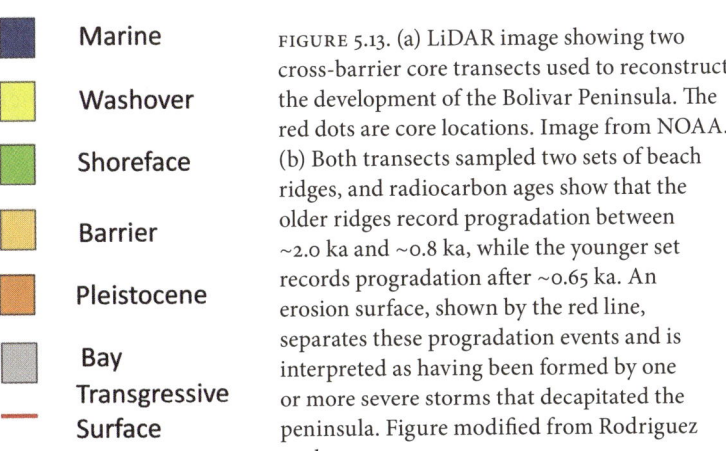

Marine

Washover

Shoreface

Barrier

Pleistocene

Bay

Transgressive
Surface

FIGURE 5.13. (a) LiDAR image showing two cross-barrier core transects used to reconstruct the development of the Bolivar Peninsula. The red dots are core locations. Image from NOAA. (b) Both transects sampled two sets of beach ridges, and radiocarbon ages show that the older ridges record progradation between ~2.0 ka and ~0.8 ka, while the younger set records progradation after ~0.65 ka. An erosion surface, shown by the red line, separates these progradation events and is interpreted as having been formed by one or more severe storms that decapitated the peninsula. Figure modified from Rodriguez et al., 2004.

Figure 5.13 (c) Photograph taken in a sand pit near core transect A. The exposures in this sand pit provided a rare opportunity to observe barrier sands resting on marsh and estuarine deposits. A storm deposit composed of sand and shells records erosion of the peninsula by one or more severe storms. Figure modified from Rodriguez et al. (2004).

in the salinity of Galveston Bay during this time in the form of a spike in fossil dinoflagellates. This microfossil is equated with more open marine conditions (Ferguson, 2018b).

Radiocarbon ages from the younger beach ridge set indicate that the western Bolivar Peninsula prograded at a rate of 2.6 to 5.1 feet/year. This barrier growth ended in historical time, when the shoreline began to erode at an average rate of 2.3 feet/year, with rates along the eastern end of the peninsula as high as 5 feet/year. The reversal from progradation to transgression is also recorded in offshore sediment cores that sampled marine mud onlapping shoreface deposits (fig. 5.13b). The exception to this history of barrier erosion is a short stretch of shoreline at the far west end of the peninsula, where sand is trapped behind a rock jetty at the mouth of the Houston/Galveston Ship Channel (fig. 5.6).

Galveston Island

Galveston Island has been a focus of research for more than six decades, making it one of Earth's most thoroughly studied barrier islands. The island decreases in thickness from 40 to 10 feet from east to west, which reflects the relief on the underlying Pleistocene surface. Prominent beach ridges dominate the island's landscape and record a long history of progradation (fig. 5.14a). The exception is the narrow west end of the island, where storm surge channels dominate the backshore landscape. Two sets of beach ridges exist. An older set is oriented toward the northeast and, similar to the Bolivar Peninsula, suggests that initial growth was relatively slow as much of the sand supply to the island ended up in the Bolivar tidal inlet and tidal delta. As the deeper portion of the inlet was filled, the shoreline orientation changed, and the progradation of the island increased. Radiocarbon ages reveal that the older ridge set spans from ~5.3 ka to ~2.6 ka and that the rate of island progradation at this time was 1.3 feet/year. The younger ridge set yielded ages between ~2.6 ka and ~1.8 ka, indicating an increase in the progradation rate to an average of 4.0 feet/year. This period of progradation is recorded offshore by the nearly 1.2-mile seaward extension of upper shoreface deposits (fig. 5.14b).

Beach ridges younger than ~1.8 ka have been eroded, indicating a more recent transgressive history for the island. This reversal in the island's development is recorded in offshore sediment cores that show approximately 0.8 mile of landward migration of the shoreface and associated onlap by marine mud (fig. 5.14b). Radiocarbon ages indicate that the landward migration of the shoreface began ~2.0 ka, followed by the landward retreat of the island. Given the slow rate of sea-level rise at this time, a decrease in offshore sand supply probably caused the retreat of the island.

Construction of the Galveston Seawall in the aftermath of the Great Storm of 1900 protected the island. Still, the construction of rock jetties on either side of the Houston/Galveston Ship Channel disrupted the longshore sand transport system. As a result, sand has accumulated at the far west end of the Bolivar Peninsula and the far east end of Galveston Island (fig. 5.6b). Erosion of the island resulted in the complete loss of the beach seaward

a

West Bay

Older Beach
Ridge Set

Younger Beach
Ridge Set

b

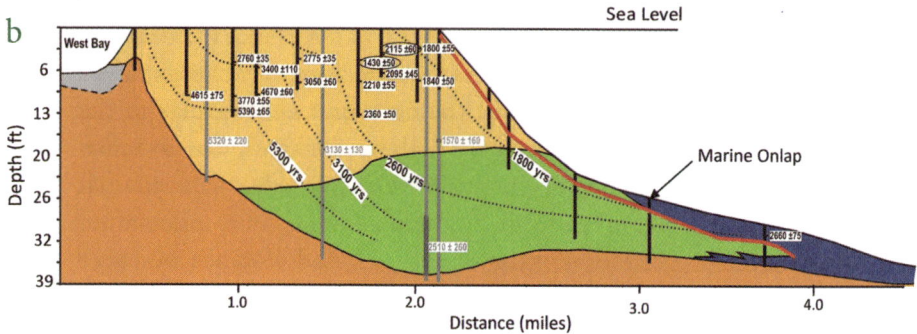

Sea Level

West Bay

Depth (ft)

6
13
20
26
32
39

2760 ±35
3400 ±110
4670 ±60
4615 ±75
3770 ±55
5390 ±65
5320 ± 220

2775 ±35
3050 ±60

2115 ±60
1430 ±50
2210 ±55
2095 ±45
1840 ±50

2360 ±50

1800 ±55

5300 yrs
3130 ± 130
3100 yrs
2600 yrs
1800 yrs
1570 ± 160
2510 ± 200
2660 ±75

Marine Onlap

1.0
2.0
3.0
4.0

Distance (miles)

MARINE

WASHOVER

SHOREFACE

BARRIER ISLAND

UNDIF. PLEISTOCENE

OPEN BAY

TRANSGRESSIVE SURFACE

FIGURE 5.14. (a) LiDAR image of Galveston Island showing two sets of beach ridges. Red dots are locations of sediment cores used to study the island's history. From NOAA. (b) Cross section through Galveston Island. Radiocarbon ages from the older ridges record slow progradation (1.3 feet/yr) between 5.6 ka and 2.6 ka, followed by more rapid progradation (4.0 feet/yr) between 2.6 ka and 1.8 ka. Younger beach ridges have been eroded, but offshore cores show marine onlap and record the more recent transgression. Modified from Rodriguez et al. (2004) and Anderson et al. (2022).

FIGURE 5.15. *Facing*, (a) Recent image of the west end of the Galveston Seawall. From Google Earth. (b) 1954 aerial photograph of the same location as above. The red lines are at approximately the same locations in both photographs. The vegetation line retreated about 400 feet between the times these photos were taken.

of the seawall by the late 1950s. Nowhere else is the rate of shoreline erosion on the island more evident than at the west end of the seawall, where
the shoreline is offset 400 feet from its 1954 location, indicating a rate of
shoreline migration of 5.7 feet/year (fig. 5.15). The historical rate of shoreline
retreat west of the Galveston Seawall varies but averages about 3 feet/year.

Most of the sand eroded from Galveston Island is transported west by
prevailing longshore currents and is deposited in the San Luis Pass tidal

delta (fig. 5.5d). Radiocarbon ages show that the tidal delta has existed for at least 3,500 years, but it was initially located east of the modern tidal delta. A study of the tidal delta's more recent evolution using historical charts and photographs showed that the delta experienced an increase in growth in historical time, attributed to an increase in erosion of Galveston Island (Wallace and Anderson, 2013).

The Central Texas Coast

Compared to the upper Texas coast, the central Texas coast has a relatively steep offshore profile and higher sand supply because of the convergence of longshore currents in the region. The result is coastal barriers that are older and thicker than east Texas barriers. These include Matagorda Island, St. Joseph Island, and Mustang Island (fig. 5.16). Detailed studies of the islands indicate that progradation and aggradation occurred between ~7.5 ka and 4.0 ka (Anderson et al., 2022). Any beach ridges formed during this early phase of progradation were eroded by a major transgressive event that began ~5.0 ka and is recorded in offshore sediment cores. With time, the islands recovered, and a new phase of progradation, recorded by beach ridges on Matagorda Island and St. Joseph Island, took place ~2.0 ka to <435 years ago. This later growth phase is recorded in offshore sediment cores that show up to 2.5 miles of shoreface progradation (fig. 5.16).

The late Holocene transgressive event that impacted central Texas barrier islands occurred around the same time that Galveston Island and the Bolivar Peninsula were prograding. This was followed by landward migration of Galveston Island and the Bolivar Peninsula around the same time that the central Texas coast began prograding. This out-of-phase barrier development is an example of the different responses of coastal barriers to changing sediment supply along the coast.

From the 1930s through 2019, Matagorda Island was generally stable. Since 2000, the island's shoreline has shifted landward at an average rate of −3.0 feet/year (Paine et al., 2021). St. Joseph Island is also retreating landward at a rate of −3.0 feet/year, and Mustang Island is retreating at an average rate of −1.0 foot/year. Thus, the central Texas coast has undergone a significant reversal from barrier progradation to landward retreat. Unlike the late Holocene differences in shoreline movement, this recent reversal in coastal migration appears to have impacted the entire Texas coast.

a

Matagorda Island

St Joseph Island

Mustang Island

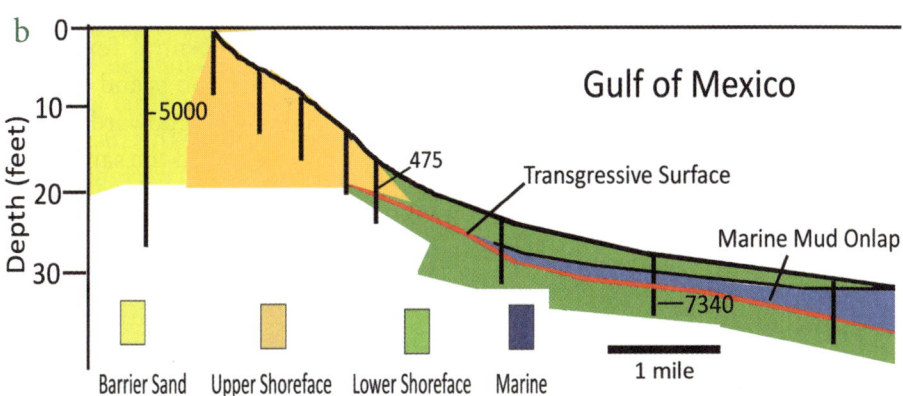

b

Depth (feet)

Gulf of Mexico

5000

475

Transgressive Surface

Marine Mud Onlap

7340

Barrier Sand   Upper Shoreface   Lower Shoreface   Marine

1 mile

FIGURE 5.16. (a) Sediment core transect (red dots) across Mustang Island and its shoreface shows a transgressive event marked by approximately 2.5 miles of marine mud onlap followed by ~2 miles of shoreface progradation. From Google Earth. (b). This section represents coastal changes recorded by multiple core transects across the central Texas shoreface. Modified from Odezulu et al. (2020).

Transgressive Barriers of Texas

Follets Island, the Matagorda Peninsula, and South Padre Island are characterized by thin barrier and shoreface sands resting on back-barrier washover and estuarine deposits, indicating a history of significant landward migration. The Brazos and Colorado delta headland separates Follets Island and the Matagorda Peninsula, and both barriers are composed of sand eroded from these deltas (fig. 4.5). South Padre Island was nourished by sand eroded from the Rio Grande delta. Of the three barriers, Follets Island has been studied in the greatest detail (Odezulu et al., 2018). It is a relatively small barrier located west of Galveston Island and separated from it by San Luis Pass. Sediment cores through the island reveal that barrier sands are generally less than six feet thick (fig. 5.17). Radiocarbon ages from washover deposits now located offshore of the island range between ~3.2 ka and ~2.7 ka, so the island was seaward of its current location during this time. The island reached its current location between ~2.4 ka and ~1.5 ka, yielding a landward migration rate between −0.1 and −0.6 foot/year. This period of relative stability resulted in the aggradation of the island, as indicated by almost vertical facies boundaries.

The sand budget for Follets Island is reasonably constrained. Little sand is transported to the island from the San Luis tidal delta, so the island's east end is rapidly eroding. Sand supply from the west was halted when the Brazos River was diverted in 1929. Historical records indicate that the rate of landward migration of Follets Island increased from −3.0 feet/year between 1850 and 2001 to the current rate of −6.0 feet/year. This increase in landward migration is resulting in an increase in the rate of washover from the island into Christmas Bay (Odezulu et al., 2018). At the current rate of erosion and washover, Follets Island will cease to exist as a subaerial barrier in about 70 years, even sooner if impacted by one or more hurricanes. Currently, portions of the island are barely wide enough to accommodate the highway connecting Surfside Beach to Galveston Island.

The Matagorda Peninsula is a long, narrow peninsula with a landscape dominated by washover features (fig. 5.7). It comprises an average of five feet of barrier sand resting on bay and washover deposits. Radiocarbon ages indicate that the peninsula was near its current location between ~3.1 ka and ~1.7 ka, so it did not migrate significantly during that time (Anderson et al., 2022). However, during historical times, the peninsula has been migrating landward at an average rate of −3.1 feet/year.

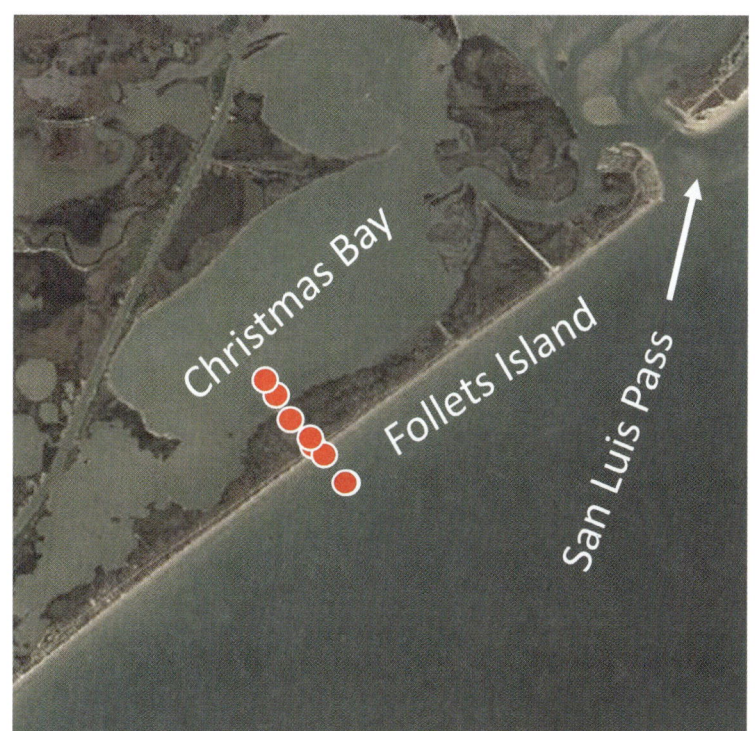

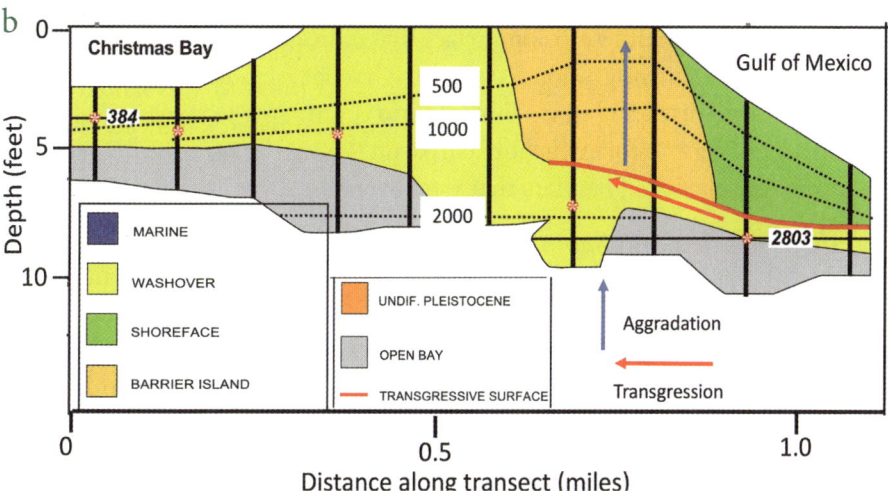

FIGURE 5.17. (a) Sediment cores collected onshore and offshore of Follets Island reveal a history of transgression during the past 3,000 years, with barrier and shoreface deposits resting on back-barrier deposits. Red dots show core locations. From Google Earth. (b) Note the aggradation of younger shoreface and barrier deposits (marked by blue arrow) that record a period of shoreline stability following transgression. Modified from Anderson et al. (2023).

Padre Island is primarily inaccessible by land and is one of the most pristine barrier islands on the Gulf Coast. Sediment cores through the island sampled up to 12 feet of barrier sands resting on washover and bay deposits, indicating a transgressive history for the island. Offshore cores sampled less than six feet of shoreface deposits resting on the red clay of the Rio Grande delta (fig. 5.18). A single radiocarbon age from the island indicates that it has existed close to its current location since ~2.1 ka, which yields an estimated long-term landward migration rate of < −1 foot/year. This location's current shoreline migration rate is more than −8.0 feet/ year, with virtually all the sand eroded from the island transported by storm washover into Laguna Madre. A climate-induced reduction in sand supply from the Rio Grande set the stage for the demise of Padre Island, but humans have also contributed through excessive water usage and dam construction (chapter 4).

Follets Island, the Matagorda Peninsula, and South Padre Island were formed by reworking the Colorado and Rio Grande deltas. Today, they are all erosion hot spots. They erode so fast that these barriers and shore-face deposits are typically less than seven feet thick and rest directly on back-barrier deposits. These back-barrier and barrier/shoreface deposits are separated by a shallow erosion surface that indicates erosion by fair-weather waves. Shoreline migration for all three barrier islands has increased significantly in historical time. This scenario is similar to what occurred in the early Holocene to form offshore banks on the east Texas continental shelf, remnants of coastal barriers that were overcome by sea-level rise.

The Calcasieu and Sabine Chenier Plains

The western Louisiana coast, from the Mississippi River Delta to Sabine Lake, is characterized by coastal wetlands and lakes separated from the Gulf by alternating mudflats and sand ridges known as chenier plains (fig. 5.2). The Sabine and Calcasieu chenier plains formed since 3.0 ka after filling the large tidal inlets that once separated Calcasieu Lake and Sabine Lake from the Gulf. The result was estuaries with restricted access to the Gulf. To the east, chenier plains occur farther inland and are interpreted as relict features that became landlocked as the eastern fringes of the delta expanded westward.

Radiocarbon and OSL ages indicate that during most of its development, the Calcasieu chenier plain prograded at an average rate of +6.6 feet/year (Hijma et al., 2017). It comprises 33 individual ridges, indicating a ridge

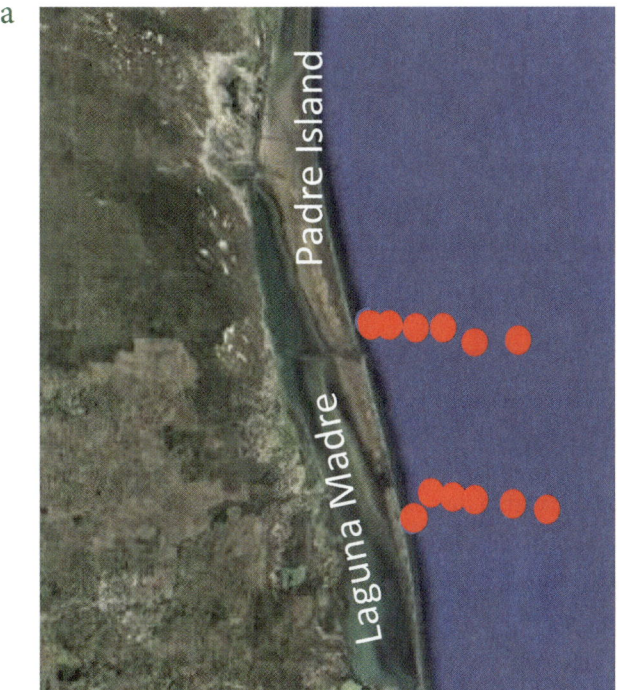

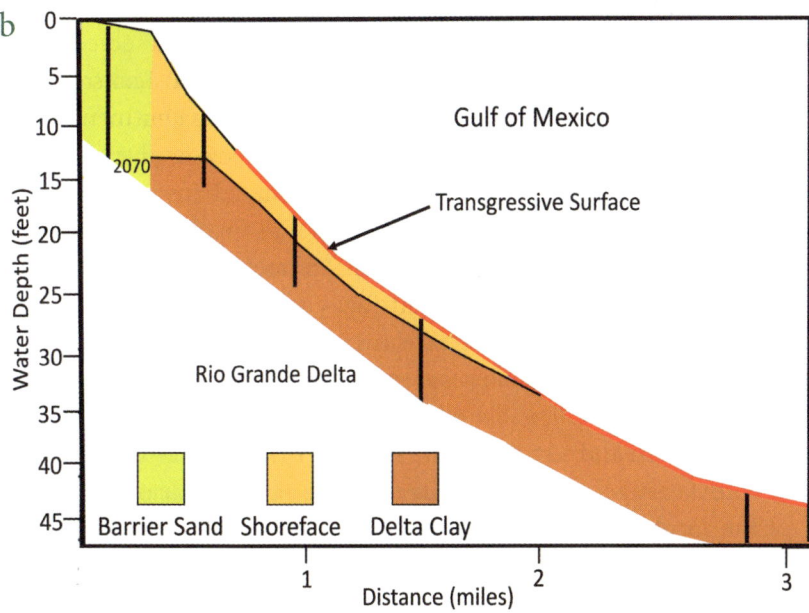

FIGURE 5.18. Sediment core profile across South Padre Island and its shoreface. (a) from Google Earth; (b) modified from Rodriguez et al. (2001).

formation rate of roughly every 90 years. This rate suggests storm-related formation, but fluctuations in fine-grained sediment output from the Mississippi River, including significant floods, have likely contributed to the formation of this unique coastal setting. The progradation rate decreased to +2.6 feet/year during the past century, with most of that growth occurring east of, and in proximity to, jetties at the entrance to the estuary. The same holds for the Sabine chenier plain.

Louisiana Barrier Islands

Several barrier islands occur offshore of Louisiana (fig. 5.19). Only Grand Isle is inhabited. These islands are an essential first line of defense for the bays and wetlands behind them.

Throughout the late Holocene, the mouth of the Mississippi River shifted in location and constructed several large delta lobes (fig. 4.1). Once these delta lobes filled the relatively shallow waters of the inner continental shelf, they were abandoned and exposed to wave erosion, forming a string of barrier islands and back-barrier bays. The Early and Late Lafourche delta lobes were reworked to form the Timbalier Islands and Grand Isle, while the Chandeleur Islands were formed by reworking the St. Bernard lobe. Abandonment and reworking of these lobes occurred between 3.0 ka and 1.0 ka, so the early stages of island formation span this time.

Traditionally, monitoring of shoreline change in coastal Louisiana was conducted by agencies and academic institutions focusing on different portions of the barrier complex and different time segments. More recently, Louisiana initiated the Barrier Island Comprehensive Monitoring Program. Relying on historical charts and aerial imagery, the program generated a map showing high coastal erosion rates along most of its barrier islands between the 1880s and 2015 (fig. 5.20). The map shows that the Timbalier Islands and Grand Isle have migrated landward at rates between −10 and −40 feet/year, with a few exceptions in areas where lateral barrier migration is occurring. The Chandeleur Islands are retreating landward at rates between −20 and −40 feet/year. The only barrier islands retreating at slower rates (average −3 feet/yr) occur along the western side of the modern delta. The rapid retreat of Louisiana's barrier islands is mainly the result of their relative thinness, high subsidence rates, and high vulnerability to storm impacts.

History has shown that the barrier islands of coastal Louisiana are especially vulnerable to hurricane impacts. The islands of the Late Lafourche

FIGURE 5.19. Aerial image showing south Louisiana barrier islands.
From Google Earth.

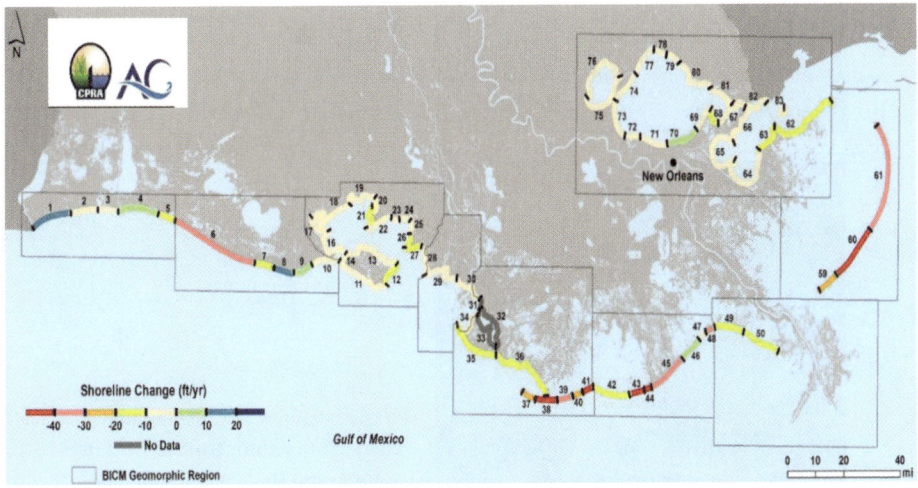

FIGURE 5.20. Louisiana shoreline change between the 1880s and 2015. From Barrier
Island Comprehensive Monitoring Program, Phase 2.

system were formed by fragmentation and detachment from a single island during a hurricane in 1853. The largest, Chandeleur Island, migrated steadily landward from 1855 until 2005, when Hurricane Katrina virtually destroyed the island, reducing it to less than 20 percent of its original surface area (fig. 5.21). Since 2005, the island has slowly recovered with the aid of sand nourishment projects but has also suffered additional storm impacts. As of 2022, the island remains low and composed mainly of storm washover deposits, so it is still highly vulnerable to storm impact.

a
Aug 7, 2005

b
Sep 8, 2005

c
2022

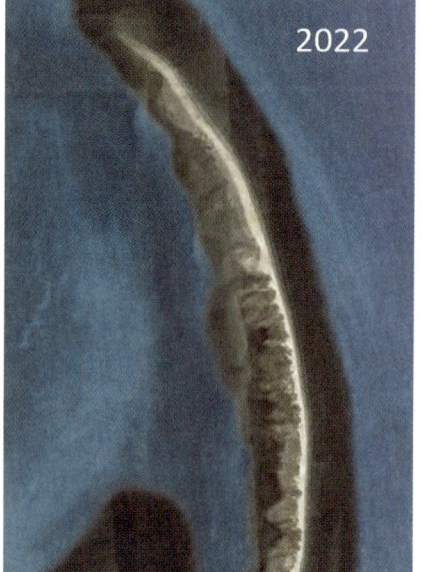

FIGURE 5.21. NASA satellite images of the northern part of Chandeleur Island. The photos from (a) August 7, 2005, and (b) September 8, 2005, were taken before and after Hurricane Katrina. (c) A recent Google Earth image showing the island after restoration.

Mississippi Barrier Islands

The Mississippi barrier chain spans the Mississippi Gulf Coast and comprises five low and narrow barrier islands (Cat, West Ship, East Ship, Horn, and Petit Bois) located between 7.5 and 12.5 miles from the mainland (fig. 5.22). Most of the islands were incorporated into the Gulf Islands National Seashore by Congress in January 1971, followed by the addition of Cat Island in March 2002. The shoreface of the islands is characterized by steep gradients that result in relatively high wave energy. Strong, westward-flowing longshore currents have contributed to the westward migration of the islands.

Geological research on Mississippi Sound and its barrier islands has been spearheaded by scientists from the University of Southern Mississippi, led for many years by Dr. Ervin Otvos, and more recently by Dr. Davin Wallace. Their research has provided a strong background for understanding how the barrier islands formed and how they are responding to current change. This work has shown that the islands are anchored on Pleistocene ridges that began to shoal ~7.5 ka when sea level reached ~ −20 feet. Remnant beach ridges occur on portions of the barriers, indicating periods of seaward progradation during their early development. This was followed by lateral accretion of the islands. By ~3.5 ka, the barrier chain had developed to the point that estuarine conditions existed in Mississippi Sound (Otvos and Giardino, 2004; Otvos, 2018).

FIGURE 5.22. Barrier islands of the west Alabama and Mississippi coast. Base map from Google Earth.

Historical charts reveal that the separation and lateral growth of the Mississippi barrier islands continued in historical time (fig. 5.23). Petit Bois Island was formed in 1740 by segmentation of the western portion of Dauphin Island during the Mobile Twin Hurricanes. Between 1849 and 1917, the island migrated just over 7.5 miles westward from Dauphin Island and eroded significantly. Horn Island also migrated dramatically westward between 1849 and 1986, and Ship Island lost about 60 percent of its surface area between 1848 and 2007. In 1969, Ship Island was breached during Hurricane Camille to form West and East Ship Islands. Cat Island lost approximately 40 percent of its surface area between 1848 and 2007 (Morton, 2007). Cat Island and West Ship Island were severely impacted by Hurricane Katrina in 2005, losing most of their unique pine and sand oak populations. Given these changes, USGS geologist Dr. Robert Morton argued that the Mississippi barrier islands are severely declining in areal extent because of erosion from both their Gulf and sound sides. Ervin Otvos has argued that Mississippi's barrier islands are among the Gulf Coast's most dynamic, fastest-changing coastal landforms. The reasons for this are clear. Extension and thinning of the islands have rendered them

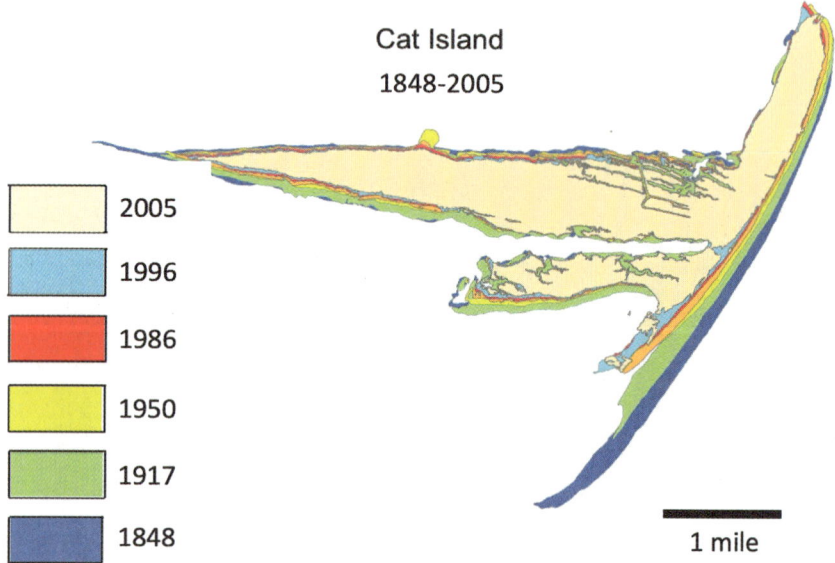

FIGURE 5.23. *Above and facing,* Beach ridge sets identified in historical charts and aerial photographs were used to reconstruct the migration history of Mississippi barrier islands. Modified from Morton (2007). Credit: US Geological Survey.

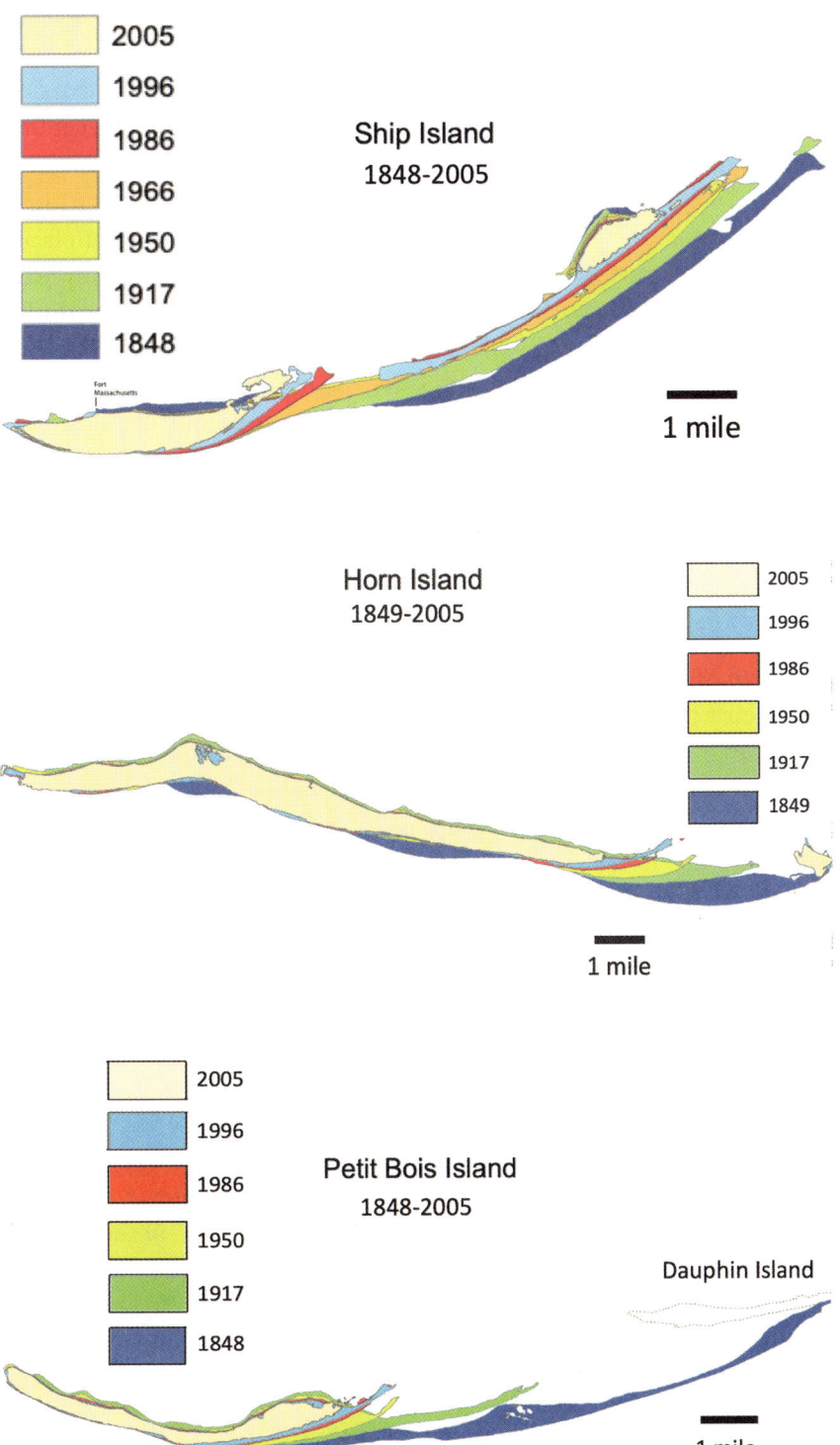

Ship Island
1848-2005

2005
1996
1986
1966
1950
1917
1848

Fort
Massachusetts

1 mile

Horn Island
1849-2005

2005
1996
1986
1950
1917
1849

1 mile

Petit Bois Island
1848-2005

2005
1996
1986
1950
1917
1848

Dauphin Island

1 mile

more vulnerable to storm overwash, which robs sand from the barriers and the longshore transport system, thus making them more susceptible to sea-level rise.

The Alabama Coast

The Alabama coast is divided into a western sector, including Dauphin Island and the Morgan Peninsula, and an eastern sector comprising a mainland shoreline with isolated lagoons (fig. 5.24). High, pine-covered dunes characterize the eastern portion of Dauphin Island, while the western portion is low and narrow, rendering it highly susceptible to storm breaching. In contrast to Dauphin Island, the Morgan Peninsula is a relatively wide and high barrier with large coastal dunes and beach ridges extending the length of the peninsula. It is a formidable barrier to storms and provides a habitat for pine forests with dense undergrowth. Sand supply to both barriers came from the western portion of the MAFLA Sheet Sand. Sand transported westward along the Morgan Peninsula ends up in the Mobile tidal delta, which has the largest ebb-tidal delta on the Gulf Coast (fig. 5.5a). Sand that makes its way to the western side of the tidal delta accumulates as a large bar, locally referred to as Sand Island, and nourishes the eastern end of Dauphin Island.

The geological histories of Dauphin Island and the Morgan Peninsula are closely tied to the evolution of Mobile Bay, which occupies the ancestral Mobile and Tensaw incised river valley, and to the presence of late Pleistocene coastal deposits that provided nucleation points for barrier growth nourished by the MAFLA Sheet Sand (fig. 5.24). Aerial photographs of the Morgan Peninsula show several sets of beach ridges with different orientations, indicating a complex formation history for the peninsula. Early development began around 5.5 ka to 5.3 ka and occurred rapidly as sand filled the shallow eastern portion of the bay. Radiocarbon and OSL ages from beach ridges on the peninsula reveal that this rapid growth occurred between ~4.0 ka and ~2.5 ka (Rodriguez and Meyer, 2006). As with the Bolivar Peninsula in Texas, the westward growth of the Morgan Peninsula slowed as it began to fill the East Mobile Valley and tidal inlet. Northwest-oriented beach ridges at the west end of the peninsula record infilling of the inlet. Likewise, Dauphin Island's thick, robust east end overlies the West Mobile Valley in contrast to the relatively thin, narrow west end.

a

b

FIGURE 5.24. (a) The west end of the Morgan Peninsula overlies the East Mobile Valley, and the east end of Dauphin Island overlaps the West Mobile Valley, which results in greater thicknesses of these barriers at these locations. From Anderson and Rodriguez (2008). DEM from NOAA. (b) Two prominent sets of beach ridges on the west end of the Morgan Peninsula record two stages of barrier evolution. An older set is oriented toward the northwest and records the infilling of a tidal inlet, and a younger set records seaward progradation of the peninsula and development of a large ebb-tidal delta. From Google Earth.

FIGURE 5.25. *Above*, Erosion at the east end of Dauphin Island has increased historically and now threatens the Civil War–era Fort Gaines, leading to beach nourishment that has restored the beach seaward of the fort. From Google Earth.

FIGURE 5.26. *Below*, Dauphin Island, Alabama, has suffered significant damage during hurricanes. The most devastating was Hurricane Katrina in 2005. The island has not regained its pre-Katrina elevation. From USGS/NASA.

USGS and state surveys conducted over the past two decades indicate that the Morgan Peninsula and Dauphin Island are rapidly eroding. The main exception is the east end of Dauphin Island, where isolated areas of beach accretion result from migration of the Mobile tidal delta. Meanwhile, the Gulf side of the island has eroded during historical times and threatens Civil War–era Fort Gaines (fig. 5.25). West Dauphin Island has an average shoreline migration rate of −5.0 feet/year and locally as high as −6.6 feet/year. This rapid erosion has been exacerbated by the dredging of the Mobile Ship Channel, which has disrupted the westward transport of sand across the Mobile tidal delta. The island's west end appears to have reached the unsustainable stage of barrier evolution where the sand supply can no longer keep pace with storm washover (fig. 5.26).

Florida's Miracle Strip

A long stretch of some of the most beautiful beaches in the world, known as the Miracle Strip, runs from Gulf Shores, Alabama, to Panama City, Florida. The sugar-textured beaches and emerald-green waters make this one of the most popular tourist attractions on the Gulf Coast. It is also one of the most highly developed coastlines of the United States.

The sands that make up the beaches of the Miracle Strip are relatively coarse and quartz rich, which maximizes the reflectivity of sunlight, resulting in snow-like brilliance. These sands came mainly from the Apalachicola, Perdido, and Escambia Rivers and their offshore deltas, which were eroded during the most recent transgression and spread across the continental shelf to form the vast MAFLA Sheet Sand (fig. 1.7).

There are few stretches of the Miracle Strip where geomorphic features remain intact; most of the landscape is covered by hotels, condos, and strip malls (fig. 5.27). Scattered mainland beaches and protected barrier islands provide a glimpse of the natural topography, including beach ridges with variable orientations that record past shoreline orientations and the locations of former tidal inlets. These features document a period of coastal stability and growth, followed by landward shoreline migration that resulted in the truncation of ridges at the modern coast. Washover features dominate the current backshore setting.

Results from a recently released (2021) assessment of shoreline change by the Florida Department of Environmental Protection characterize most of the Miracle Strip as a "critically eroding" shoreline. This widespread erosion reflects a significant shift from coastal stability and growth during

FIGURE 5.27. Aerial image showing the contrast between undeveloped Santa Rosa Island and heavily developed Okaloosa Island to the east. Santa Rosa Island is part of the Gulf Islands National Seashore and has the region's longest stretch of undeveloped shoreline. From Google Earth.

the late Holocene, attributed to the historical acceleration of sea-level rise (Anderson et al., 2023; Parkinson and Wdowinski, 2023).

The Apalachicola Delta Headland

East of Panama City, the coastline is lined by barrier islands, including Dog Island, St. George Island, St. Vincent Island, and the St. Joseph Peninsula (fig. 5.28). These islands have been sparingly altered by development and retain many of their spectacular beach ridges and expansive dunes. Their formation was closely tied to the landward retreat and erosion of the Apalachicola delta headland, which was a more formidable delta during the late Pleistocene and early Holocene (fig. 1.5). Bathymetric images collected offshore of the islands show sand ridges believed to be reworked remnants of former shorelines (fig. 5.29).

St. Vincent Island was the site of a geological investigation by two geologists from McMaster University (Rink and Lopez, 2010). Their OSL ages from beach ridges on the island record up to 3.7 miles of seaward growth since ~2.8 ka. The island prograded at an average rate of 8.8 feet/year during its early development. This was followed by even faster progradation (16.0 feet/yr) between 2.8 ka and 1.9 ka, followed by a decreased rate of 2.3 feet/year after ~1.9 ka (fig. 5.30). Historical records reveal that St. Vincent Island was still prograding at +1.6 feet/year between 1855 and 1998. Since 1998, the island has been migrating landward at an average rate of −1.2 feet/year, with some portions as fast as −10.0 feet/year.

FIGURE 5.28. Aerial image showing coastal barriers formed by reworking of the ancestral Apalachicola delta. From Google Earth.

FIGURE 5.29. Bathymetric map of the continental shelf offshore of Apalachicola, Florida, showing offshore sand ridges interpreted as remnants of paleoshorelines. From GeoMapApp.

FIGURE 5.30. Image of St. Vincent Island showing core locations and OSL ages from beach ridges. From Google Earth; data from López and Rink (2008).

The Southwest Florida Coast

From Tampa Bay to Naples, Florida, the southwest Florida coast comprises narrow mainland beaches and barrier islands, including Gasparilla Island, North Captiva Island, South Captiva Island, and Sanibel Island (fig. 5.31). These islands are believed to be composed mainly of sand delivered to the coast by the Caloosahatchee and Peace Rivers during sea-level lowstands and then reworked and transported onshore during the most recent transgression.

The coast between Gasparilla Island and Sanibel Island has seen only moderate development. It is one of the few portions of this stretch of coast where beach ridges are relatively intact. These barrier islands were the focus of an investigation by Dr. Frank Stapor and colleagues, who acquired sediment cores and radiocarbon ages needed to investigate their geological history. The results showed that the initial growth of the islands began at ~3.0 ka and was interrupted by brief episodes of erosion (Stapor et al., 1991). The results from Sanibel Island, which is representative of other islands in this chain, indicate that ~3 miles of seaward growth occurred during the past 2,000 years, yielding a progradation rate of ~8.2 feet/year. Historical charts and aerial photographs of Sanibel Island show that its shoreline was

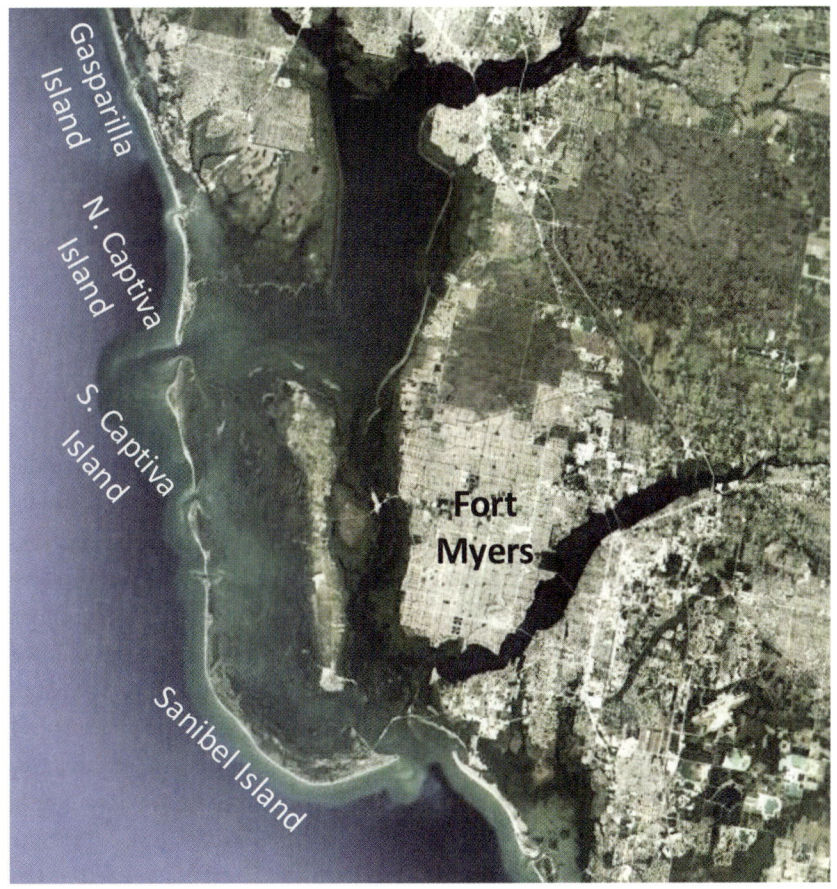

FIGURE 5.31. Barrier islands of the southwest Florida coast. From Google Earth.

advancing seaward at an average rate of +3.6 feet/year between 1855 and 2000. Since 2000, the progradation rate has slowed to +1.9 feet/year. So, while it is one of the few prograding Gulf Coast barrier islands, its rate of seaward growth has decreased in the past two decades.

Late Holocene Variability in Barrier Evolution Controlled by Sand Supply

Most of the modern Gulf coastal barriers and the Calcasieu and Sabine chenier plains formed during the late Holocene when sea level was near its current elevation, and the average rate of rise was only 0.2 foot/century (0.5 mm/yr) (fig. 1.4b). While these were ideal conditions for barrier and chenier formation, the timing and rate of their development varied across the Gulf Coast because of regional variation in sand supply. Sand came from rivers and offshore sources, the latter of which is believed to have been dominant. West Florida and Alabama rivers all flow into bays that have existed throughout the Holocene. Thus, direct sand supply from rivers to this portion of the coast has been minimal. Instead, most of the sand that composes the eastern Gulf Coast came from erosion of the offshore Apalachicola, Perdido-Escambia, and Mobile deltas (fig. 1.5). Most of that sand still resides on the continental shelf in the form of the vast MAFLA Sheet Sand (fig. 1.7).

Erosion of large lobes of the Mississippi River Delta yielded most of the sand that makes up the barrier islands of Mississippi and Louisiana (fig. 4.1). Mississippi and eastern Louisiana barrier islands in part came from erosion of the St. Bernard lobe of the delta. In contrast, erosion of the western lobes of the delta provided sand to barrier islands west of the modern delta. In Texas, the Brazos, Colorado, and Rio Grande delivered enough sand to the coast during the early and middle Holocene to construct large deltas. By late Holocene time, these deltas were reduced to wave-dominated deltas as sediment supply from these rivers declined. Other Texas rivers flowed into estuaries throughout the Holocene, so sand from these rivers did not reach the coast. Thus, erosion of the offshore Brazos, Colorado, and Rio Grande deltas produced most of the sand that composes Texas coastal barriers. With time, the landward-migrating shoreline bypassed these sand sources and they were buried in marine mud, leaving the coast undernourished with sand. Ironically, the coastal barriers closest to these rivers, namely Follets Island, the Matagorda Peninsula, and South Padre Island, have the fastest erosion rate on the Texas coast. This further highlights the minor role of rivers in supplying sand directly to the coast today.

Central Texas barrier islands, including Matagorda Island, St. Joseph Island, and Mustang Island, are Texas's oldest and thickest barrier islands. Most of the sand that makes up these islands was eroded from the offshore Colorado and Rio Grande deltas and transported by longshore currents flowing from the south and northeast. As the shoreline migrated landward during the late Holocene, these offshore sand sources were buried beneath the Texas Mud Blanket, which expanded significantly during the late Holocene, resulting in a pause in progradation for all three islands. This pause ended ~1 ka when these islands started to prograde. At the same time barrier islands of central Texas were prograding, the Bolivar Peninsula and Galveston Island were transgressing (figs. 5.13 and 5.14). Differences in sand supply were responsible for this variable behavior.

The Demise of Transgressive Barriers

In general, shoreline progradation results in broader and higher coastal barriers that are more resilient to sea-level rise and storm impact, while transgressive barriers become narrower and lose elevation over time. Thus, transgressive barriers become more susceptible to storm overwash and ultimately reach a point where the amount of sand removed by washover equals or exceeds the amount of new sand delivered to the barrier. This is called the rollover state of barrier evolution and marks the final stage in a barrier's lifespan.

Mexico's Tamaulipan coastline provides a visual perspective for a coast in the rollover stage. There, the arid climate results in minimal sand supply to the coast, and coastal barriers cannot fully recover from severe storms. The result is a coastline dominated by storm washover features and a coastal lagoon, Mexico's Laguna Madre, that frequently changes salinity (fig. 5.32). This is the destiny of Texas's South Padre Island and Laguna Madre. It is also the destiny of Follets Island and the Matagorda Peninsula in Texas and Dauphin Island, Alabama, all experiencing high coastal erosion and overwash rates. Furthermore, these barriers are only a few feet thick and subject to decapitation by storm waves. Given the meager supply of sand to these barriers, their demise is inevitable.

Offshore of Florida and Alabama, Holocene deposits are mostly re-worked or covered by the MAFLA Sheet Sand. Hence, the record of coastal evolution comes from onshore sources, and that record is limited to a few locations where the coastal geomorphology has not been overprinted by development. Studies along the eastern Gulf Coast have revealed a reversal from barrier growth in the late Holocene to slow growth or erosion during historical times. Most of the Florida Gulf Coast is currently experiencing "critical" erosion (Dean and Houston, 2016). The same is probably true for the Alabama coast, but records of historical shoreline change are insufficient to determine the timing and magnitude of change. Mississippi barrier islands have well-established histories of late Holocene growth that have shifted to dramatic lateral migration and barrier erosion during historical

FIGURE 5.32. (a) Image of a coastal barrier on the north-east Mexican coast. (b) This stretch of the coast is exposed to frequent overwash from tropical storms and cannot gain sufficient elevation to slow this process. From Google Earth.

times. Given their finite sand supply, high subsidence rates, and constant impact from severe storms, Louisiana barrier islands have always struggled to survive. With the rapid acceleration in sea-level rise, they are even more vulnerable now. Texas coastal barriers have yielded detailed geological records showing highly variable development during the late Holocene, but evidence points to a reversal to landward shoreline migration during historical time.

The shift toward landward shoreline migration across the Gulf Coast indicates that while sea-level rise played a secondary role in coastal evolution during the late Holocene, it is reclaiming its dominant role in coastal evolution (Anderson et al., 2023). Humans have contributed to the observed historical changes in Gulf Coast shorelines. Direct human impacts include dam construction and alteration of river sediment transport pathways, urbanization, dredging of navigation channels and construction of associated rock jetties, and construction of other hard structures that, ironically, are intended to combat shoreline erosion. Virtually every river that flows directly into the Gulf of Mexico has been altered in one way or another. There are no dams on the Mississippi River, but there have been significant alterations to its drainage and sediment pathways, including "maintenance" of the river mouth. These projects have significantly impacted south Louisiana, but their impact on barrier islands has been minor compared to hurricane impacts.

The Rio Grande is experiencing increasing drought conditions exacerbated by excessive water usage and dam construction in the lower portion of the river, which have had a combined impact on water and sediment discharge. However, direct sand supply from the river has been decreasing throughout the Holocene, so it is hard to gauge the relative impacts of natural and anthropogenic changes. The 1929 diversion of the Brazos River mouth by the US Army Corps of Engineers clearly influenced where sediment was being delivered to the coast. That effect was mainly in the form of a change in the location of the Brazos delta and was confined to within a 12-mile stretch of the coast. In addition to human impacts on the coast, hurricanes have also taken a toll, and that toll will increase in the future as warmer Gulf waters fuel bigger, stronger, and wetter storms. Still, the Gulf-wide shift toward coastal erosion is a red flag signaling the renewed dominance of sea-level rise in controlling shoreline behavior across the Gulf Coast.

The most effective and environmentally friendly way to combat coastal erosion is beach nourishment. Florida and Alabama have ample offshore sand resources for nourishment that are located near the coast. Mississippi barrier islands are also located near extensive sand deposits within Mississippi Sound and on the inner continental shelf (fig. 1.6b). However, these states need coastal sand budget analyses to determine how much sand is required and where to obtain that sand. Louisiana has a comprehensive, coast-wide sand budget and resource analysis program, but saving most of the state's coastal barriers may not be possible. In Texas, where the coast has highly variable shoreline erosion rates, there are good records of where and how fast the coast erodes. Still, there are few quantitative sand budget analysis results for determining how much sand is needed and where beach nourishment can be most effective. Also, more work is required to locate and conduct volume estimates of offshore sand bodies. We know nearshore sand resources are limited, including old river channels buried by thick marine mud. However, large deltas on the continental shelf are known to have wide and extensive sand bodies (Anderson et al., 2004, 2022). Unfortunately, these are over 20 miles from shore and will be challenging and costly to exploit.

Wave-dominated coastlines across the Gulf Coast vary in origin and character, but they have one thing in common. The vast majority are migrating landward. This is a reversal from the late Holocene when most coastlines prograded or were relatively stationary. This reversal is mainly the result of accelerated sea-level rise and human alteration of the coastal sediment supply and dispersal system. Thus, humans are primarily responsible for ongoing coastal erosion and should bear responsibility for preserving the coast. The most effective and environmentally conducive means of combating shoreline erosion is beach nourishment using offshore sand resources. Sand resources are relatively abundant offshore of Florida, Alabama, and Mississippi. Nearshore sand resources are more isolated in Louisiana and Texas, and most are buried in marine mud. Large deltaic sand bodies are far offshore and will be costly to exploit. Given current rates of shoreline migration, it is essential to identify and assess these sand resources.

6

Estuaries

Estuaries of the Gulf Coast have undergone a remarkable transition during the past few thousand years as the barrier islands and peninsulas that isolate them from the Gulf of Mexico evolved, gradually restricting inflow from the Gulf. The net effect has been to render them more sensitive to climate-induced changes in freshwater inflow from rivers and streams, especially in Texas, where more arid conditions result in greater variation in freshwater input. Larger estuaries have extensive bayhead deltas that rising sea levels will dramatically impact during the next few decades. Background map from Google Earth.

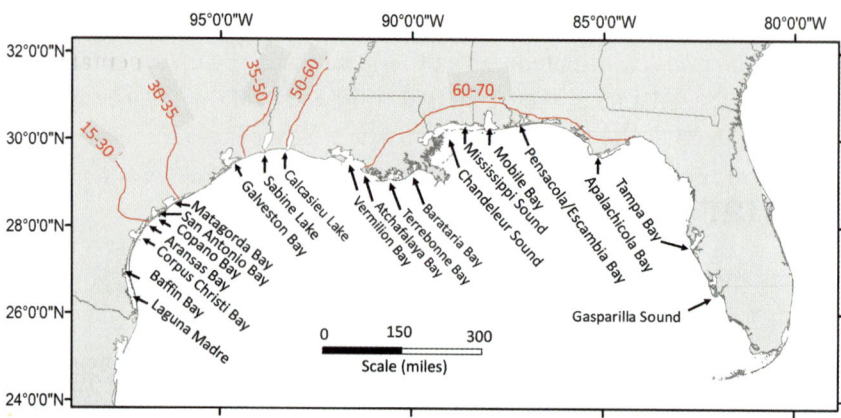

FIGURE 6.1. Locations of Gulf Coast estuaries mentioned in the text. Red lines designate mean annual precipitation in inches/year.

An estuary is a partially enclosed water body with one or more rivers or streams flowing into it and one or more inlets connecting it to the open sea. Estuaries provide an essential filtering system for freshwater runoff, are nursing grounds for many marine species, store huge quantities of carbon, and provide boating, fishing, and ecotourism opportunities. Gulf Coast estuaries are among the most vulnerable coastal environments to sea-level rise and climate change. This vulnerability has been exacerbated by human encroachment and alteration of their bathymetry and connectivity to the Gulf of Mexico by the construction of navigation channels.

The estuaries of the Gulf Coast vary significantly in size, shape, and climate setting (figs. 6.1 and 6.2). They fall into two categories based on their formation and connection to rivers. The largest estuaries were formed by submersion of river valleys during sea-level rise and remain connected to rivers. Those in the second category were formed by inundation of low areas landward of coastal barriers with little or no river connectivity. They vary widely in their connectivity to the Gulf of Mexico, resulting in very different salinity regimes, ranging from hypersaline estuaries to freshwater lagoons.

River-Influenced Estuaries

During the last interglacial (~120 ka), rivers that flowed across Louisiana and Texas occupied broad channel belts similar to those in the region today. For the next ~100,000 years, sea levels fell, causing rivers and streams to cut deep valleys that extended across the continental shelf. This period of *fluvial incision* culminated ~22 ka to ~18 ka when the continental shelf was fully exposed and dissected by river valleys that were, on average, 130 to 150 feet deep at the location of the modern coast (fig. 1.6). During the subsequent sea level rise, these valleys were partially filled with sediment and ultimately flooded to form river-influenced estuaries.

a

b

FIGURE 6.2. (a)–(b) Images illustrating variability in Gulf Coast estuaries. From Google Earth.

c

FIGURE 6.2. (c)–(g) *Left and facing*, Images illustrating variability in Gulf Coast estuaries. From Google Earth.

d

e

f

Sabine
Lake

g

Escambia
Bay

East Bay

Pensacola Bay

West Florida is characterized by relatively deep, narrow estuaries that retain the general shape of the river valleys they occupy. This shape reflects the higher coastal plain relief in the region, which was not conducive to river meandering and channel widening. The Pensacola Bay/East Bay/Escambia Bay complex provides a good example (fig. 6.2g). Branching estuaries are also associated with smaller rivers and streams. For example, Baffin Bay in south Texas occupies a branching river valley cut by streams with low discharge (fig. 6.2b). Mobile Bay is a broad, north–south-oriented estuary that occupies a relatively deep valley cut by the Alabama, Tombigbee, and Tensaw Rivers (fig. 6.2e). The combined discharge of these rivers was high enough to cut a relatively straight path through the region's generally steep coastal plain relief.

The estuaries of western Louisiana and Texas vary widely in shape, but most have rounded shapes that resulted from inundation of their shallow valley margins. Galveston Bay provides a good example, with rounded shorelines that are flooded river meanders (fig. 6.2c). Calcasieu Lake and Sabine Lake occupy similar valleys, but the shallow margins of these valleys have been filled with sediment and are now occupied by wetlands (fig. 6.2f). Mississippi Sound inherits its unique configuration from the relatively steep, sandy coastal plain shaped by a network of branching river and tributary valleys, the largest being the Biloxi and Pascagoula valleys (figs. 1.6b and 6.2d). The linear chain of barrier islands that separate Mississippi Sound from the Gulf of Mexico was formed relatively far offshore of the main coastline by reworking of deltaic deposits of the Mississippi Delta and transgressive ravinement of offshore sand sources. Apalachicola Bay also occupies an area of high relief and sandy substrates. Its deep, narrow valley is filled mainly with Holocene sediments. As with Mississippi Sound, its shape resulted from the formation of a chain of barrier islands by wave reworking of the ancestral Apalachicola delta (fig. 6.2a).

The second category of estuaries and lagoons are shore-parallel, shallow water bodies formed by the flooding of low-lying areas behind coastal barriers (fig. 5.1b). They extend along most of the coast of Texas from East Bay to the Rio Grande. In Florida, they occur mainly along a stretch of coast between Santa Rosa Sound and Apalachicola. Landward migration of these water bodies is mostly restricted by remnants of the Pleistocene Ingleside Paleoshoreline, with elevations that exceed +8 feet in most areas (fig. 6.3). Elsewhere, spoil banks of the Intracoastal Waterway block their landward migration. As a result, they are being filled with washover deposits.

Corpus Christi Bay

Aransas Bay

Matagorda Bay

Ingleside Paleoshoreline

Pleistocene Beach Ridges

Saint Andrews Sound

Holocene Beach Ridges

FIGURE 6.3. The inland shoreline of most shore-parallel estuaries occurs at the topographic break associated with the Pleistocene Ingleside Paleoshoreline and by spoil ridges of the Intracoastal Waterway. As a result, these estuaries are restricted in their ability to migrate landward in response to sea-level rise. From Google Earth.

Estuarine Circulation, Salinity, and Sedimentation

Because estuaries vary widely in shape and bathymetry, they have different wind- and tide-driven circulation. Their salinity regime is governed mainly by river and stream inflow and tidal exchange with the Gulf of Mexico. The latter is controlled by the number, size, and depth of tidal inlets that connect these estuaries to the Gulf. Freshwater inflow to estuaries varies widely across the Gulf Coast and is strongly influenced by climate. In south Texas, Laguna Madre and Baffin Bay occur in an arid climate setting, and both estuaries are characterized by low freshwater inflow and by restricted connections to the Gulf of Mexico (fig. 6.1). As a result, both are hypersaline. In contrast, Sabine Lake and Calcasieu Lake experience more humid conditions. Both estuaries receive relatively large freshwater inflow and have restricted tidal inlets. The result is relatively low salinities, but neither is actually a lake. Other estuaries, such as Galveston Bay and Mobile Bay, are characterized by relatively high saltwater inflows from the Gulf. Still, their freshwater inflow varies seasonally and results in seasonal changes in salinity.

Most Gulf Coast estuaries are tide dominated, meaning that tides dominate their circulation. During rising (flood) tides, salty water from the Gulf flows into estuaries, and during falling (ebb) tides, outgoing flow flushes them. The efficiency of this process is determined by the shape of the estuary and the size of its tidal inlet(s). In large estuaries, where tidal interaction with the Gulf is through one or more large inlets, circulation is generally characterized by multiple cyclonic cells that distribute salty water during rising tides (fig. 6.4a). The outgoing flow is usually more of a direct flushing process that delivers sediment and nutrient-laden bay waters to the Gulf (fig. 6.4b).

Tidal circulation in Tampa Bay is more complicated than that of Mobile Bay mainly because of its irregular shape and bathymetry (fig. 6.5). Freshwater inflow to Tampa Bay is significantly lower than that of Mobile Bay, promoting higher salinities. One would think that this would equate to higher water quality in Tampa Bay. However, urban development and restricted tidal flushing have resulted in generally poor water quality for the estuary and the accumulation of smelly bottom sediments that the locals refer to as "muck."

Most sediment delivered to the Gulf Coast by smaller rivers and streams is deposited in estuaries. This sediment is vital to maintaining the fringing

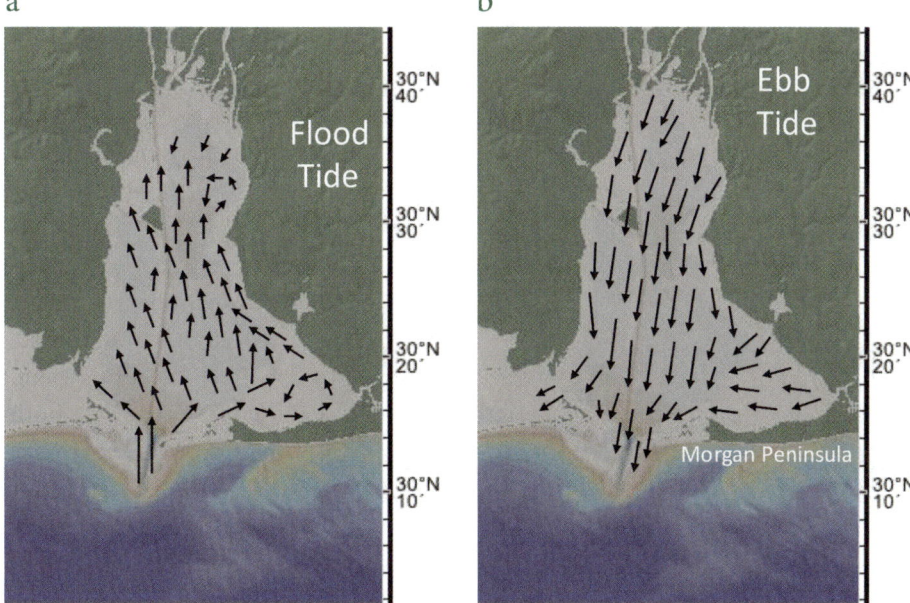

a

Flood
Tide

30°N
40´

30°N
30´

30°N
20´

30°N
10´

88°W 10´   88°W   87°W 50´

b

Ebb
Tide

Morgan Peninsula

30°N
40´

30°N
30´

30°N
20´

30°N
10´

88°W 10´   88°W   87°W 50´

c

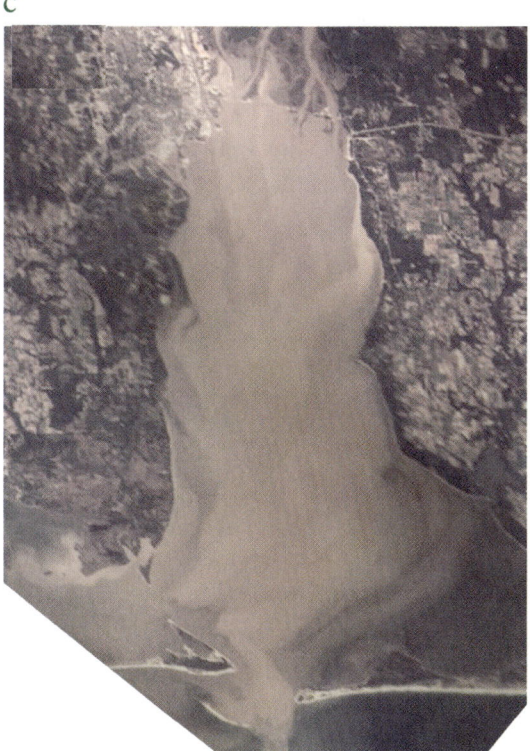

FIGURE 6.4. Generalized tidal circulation in Mobile Bay, Alabama, during (a) flood and (b) ebb tide. Background from GeoMapApp. (c) Photo taken during a high discharge event shows suspended sediments being flushed from the estuary during an ebb tide. NASA photo.

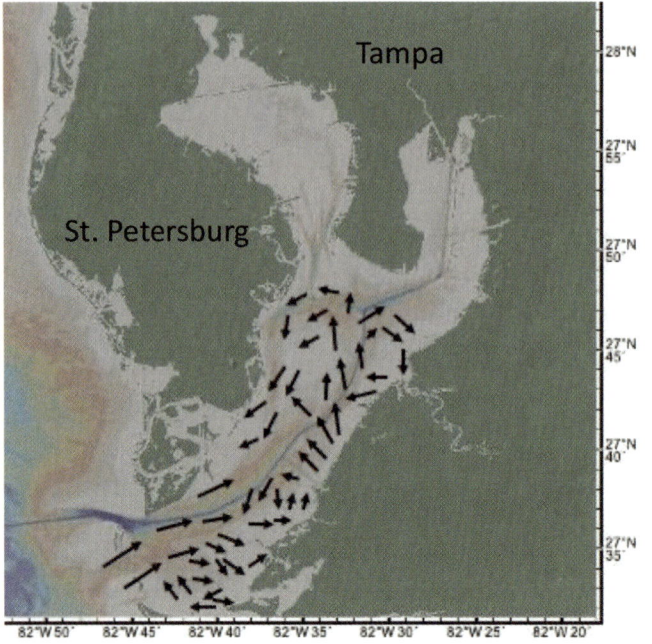

FIGURE 6.5. Generalized tidal circulation model for the southern portion of Tampa Bay illustrating complex flow patterns resulting from the estuary's irregular shape and bathymetry. Based on a circulation model by Weisberg and Zheng (2006); base map from GeoMapApp.

wetlands and bayhead deltas within these estuaries, which are significant components of their ecosystems. Estuaries from east Texas to northwest Florida experience high precipitation and relatively high sediment discharge. Estuaries within more arid regions, such as the south Texas coast, are characterized by low water and sediment discharge, resulting in wetlands that struggle to keep pace with rising sea levels. Rainfall along the semiarid central Texas coast is sufficient to maintain relatively high water and sediment discharge. However, climate models predict more arid conditions in the region that could result in a northward shift toward hypersaline estuary conditions.

Lessons from the Past

Given the vulnerability of estuaries to climate change and sea-level rise, we need to assess what can be learned from their response to changes during the Holocene, especially the early Holocene when rates of sea-level rise were similar to current rates and the IPCC's projected rates (fig. 2.21). Prompted by this question, a long-term research program was started by scientists from Rice University, the University of Alabama, and the University of

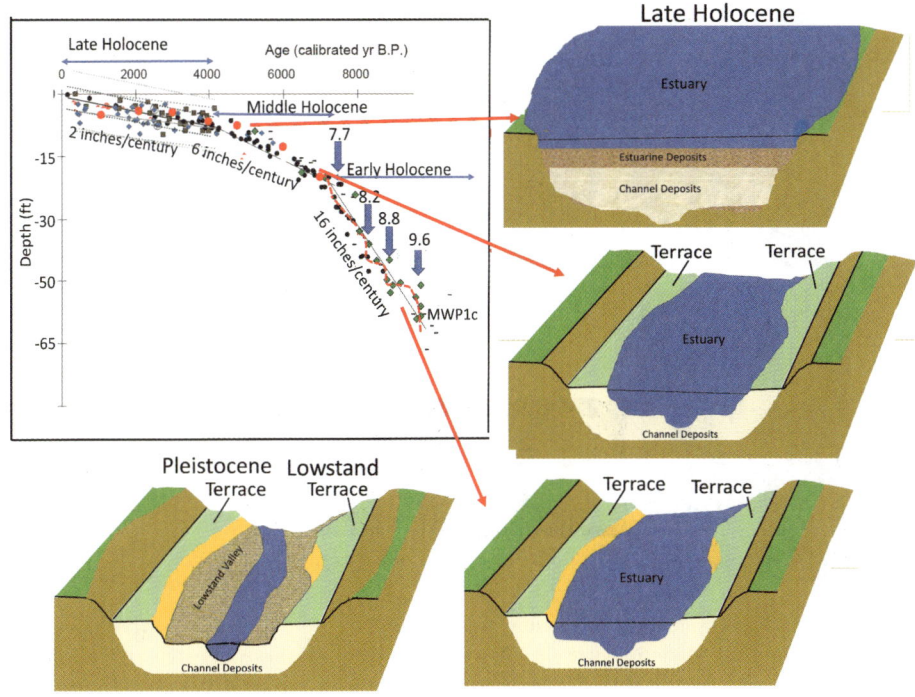

FIGURE 6.6. Model illustrating the evolution of river-influenced estuaries during sea-level rise. The narrow, deeper portions of most paleovalleys were flooded during the early Holocene, resulting in long, narrow estuaries. As the rate of sea-level rise decreased, the valleys were partially filled with sediment, and their broad margins (former floodplains) were flooded to form wide, shallow estuaries. Sea-level data modified from Anderson et al. (2022).

California, Santa Barbara. The study spanned a decade and included Mobile Bay, Calcasieu Lake, Sabine Lake, Galveston Bay, Matagorda Bay, Copano Bay, Corpus Christi Bay, and Baffin Bay. A similar methodology was used for each site and involved two vessels. One vessel was dedicated to conducting seismic surveys. The other vessel was equipped with a water well–type drilling rig and used to acquire sediment cores.

The model in figure 6.6 illustrates a river valley's expected flooding history and an estuary's formation. Note that the valley has terraces formed when sea level was relatively stationary. The final stage in valley sculpting occurs during the *lowstand* when a narrow, deep valley is cut. Initial sea-level rise inundates this deep incision to form a narrow, elongated estuary. With continued sea-level rise, the terraces are flooded, resulting

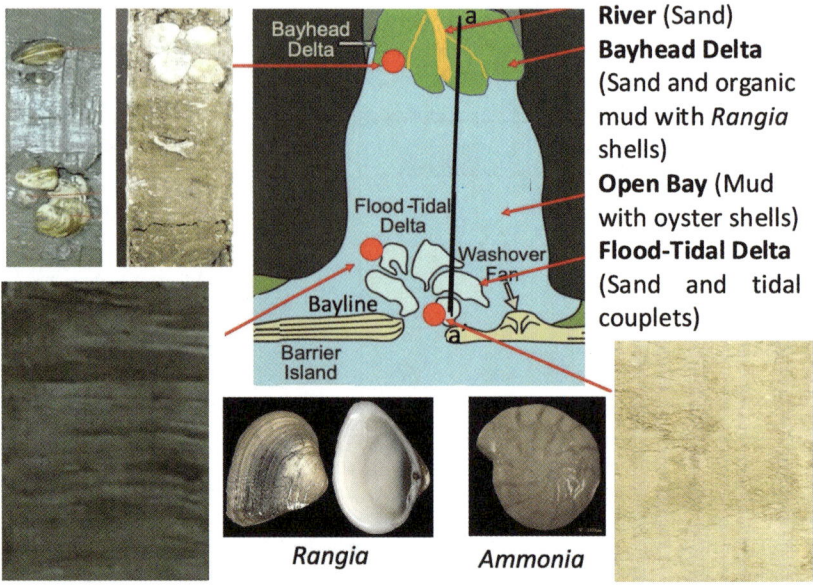

FIGURE 6.7. Sedimentary environments and facies of a typical river-influenced estuary. Profile a-a' as shown in figure 6.8. Photos of core sections whose locations are indicated with red dots are also shown. Photos at the upper left are typical bayhead delta core sections of organic mud (left) and interbedded mud and sand (right) with abundant *Rangia* shells. The photo at the lower left is of a core section with layers of mud and sand that are typical of the fringes of a flood-tidal delta. The photo at the bottom right shows a core section with the well-sorted sand of a proximal tidal delta. Also shown is a photo of *Rangia* shell halves and a magnified image of *Ammonia*, a foraminiferan diagnostic of upper and middle bay environments.

in higher inundation rates and widening of the estuary. Broad, low terraces remain a significant component of the existing fluvial topography of several Gulf Coast estuaries, so they are projected to continue to flood in the coming decades.

As is the case for coastal barriers and chenier plains, analysis of the response of estuaries to sea-level rise and climate change relies on an understanding of sedimentary facies associated with different estuarine environments and how their stratigraphic stacking patterns can be used to interpret changes in these environments. Figure 6.7 illustrates a typical river-influenced estuary and its main habitats. From the river mouth to offshore, these include river, bayhead delta, open bay, and lower bay tidal inlet and tidal delta environments. River deposits comprise mainly sand and gravel, with rare freshwater mollusk shells. These fluvial sands transition into bayhead delta deposits, including interbedded sand and clay with the shells of

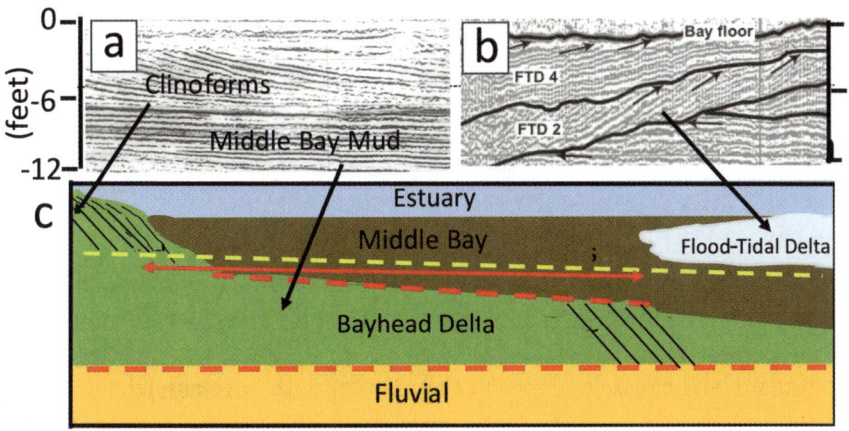

FIGURE 6.8. Examples of estuarine seismic facies include (a) seaward-dipping beds (clinoforms) that record progradation of a bayhead delta into the middle bay and (b) high-angle cross-beds formed by lateral migration of a tidal inlet. (c) Illustration of stratigraphic surfaces used to interpret estuarine evolution. Units above the yellow dashed line record the progradation of the bayhead delta and the formation of the tidal delta, indicating a period of stability and growth of these estuarine environments. The red dashed lines designate flooding surfaces separating more seaward (above) from landward (below) environments. The red arrow shows the magnitude of flooding indicated by landward shifts in the bayhead delta clinoforms.

Rangia, a mollusk inhabiting the delta fringe environment, and mud with abundant organic material. The open bay environment is characterized by mud with oyster shells and includes microfossils (foraminifera and *ostracods*) that are relatively common in estuarine sediments. These microfossils are good paleosalinity indicators and have even been used to reconstruct salinity changes through time. Finally, the lower estuary includes the flood-tidal delta, which is composed of sand and alternating thin-bedded sand and mud (tidal couplets) with a variety of mollusks and microfossils that are diagnostic of tidal delta and shallow bay environments.

The first step in analyzing an estuary's evolution is acquiring seismic data to map the valley topography and identify seismic units and surfaces required to reconstruct changes in estuarine environments. The USGS acquired most of the seismic data for mapping the valleys during the late 1970s. Later seismic surveys by Rice University and the University of Alabama involved methods capable of resolving sedimentary layers only a few feet thick. These high-resolution seismic data are used to recognize *seismic facies* that are unique to specific estuarine environments (fig. 6.8). Figure 6.8a shows an example of seaward-dipping strata (*clinoforms*) formed by progradation of a bayhead delta. These clinoforms overlie parallel layered strata that are diagnostic of the middle-bay environment. Figure 6.8b is a

segment of a seismic profile collected near the mouth of an estuary showing large-scale cross-beds formed during lateral migration of a tidal inlet.
When the same stratigraphic surfaces bound the bayhead delta, middle bay,
and tidal delta deposits, they record a period of estuarine stability when
the bayhead delta was prograding, the middle estuary was filling with
mud, and a large tidal inlet existed (fig. 6.8c). A landward shift in all three
of these facies indicates a flooding event, which is marked by a flooding
surface. Figure 6.8c illustrates two flooding surfaces, one with bay mud
resting on bayhead delta deposits and another with bayhead delta deposits
resting on fluvial deposits. Both surfaces indicate significant landward
shifts in these environments. The magnitude of flooding events is determined by measuring the distance between preflooding and postflooding
facies, such as bayhead delta clinoforms, using seismic profiles collected
along the axis of the bay. The combined results from seismic surveys and
sediment core analyses record how individual estuaries evolved. The final
step in reconstructing estuary evolution involves radiocarbon dating to
establish the timing and rates of change. An important assumption is that
flooding events caused by a rapid rise in sea level would be widespread and
impact multiple estuaries. In contrast, more localized flooding surfaces are
more likely to have been produced by reduced sediment supply to estuaries
caused by climate change.

The Holocene Evolution of Gulf Coast Estuaries

Early Holocene Response to Punctuated Sea-Level Rise

Since current rates of sea-level rise are in the range of those during the early
Holocene, that period should provide insight into how Gulf Coast estuaries
responded to faster rates of rise, including times of punctuated sea-level
rise. Studies on the east Texas continental shelf during the late 1960s and
1970s provided subtle evidence that the shoreline shifted landward in a
stepwise fashion during the early Holocene. This research was followed
by detailed seismic surveys on the continental shelf that provided the data
needed to map offshore river valleys (fig. 1.6a). Combined with the Holocene
sea-level curve (fig. 1.4b), these data have yielded a series of paleogeographic
maps depicting the history of paleovalley flooding and estuary formation
(fig. 1.9). These maps show dramatic changes in the coastline and estuaries
during the Holocene. Sediment cores reveal that the Brazos and Colorado
valleys are filled with river deposits, indicating that sediment supply from

these rivers was high enough to keep pace with sea-level rise, even during the early Holocene (Taha and Anderson, 2008).

The issue of punctuated sea-level rise and impacts on estuarine evolution was the subject of research focused on the offshore Trinity and Sabine valleys during the late 1980s (Thomas and Anderson, 1989). These valleys were found to contain massive tidal inlet deposits separated by over five miles along the valley axis (fig. 1.8). This indicates that the shoreline retreated landward in a stepwise fashion and then paused long enough for the formation of new coastal barriers and tidal inlets. Seismic data were also used to map the locations of bayhead deltas, which show a history of landward stepping followed by stability and growth (fig. 6.8). One important observation was that tidal inlets and bayhead deltas that occur at the same stratigraphic level are bounded by flooding surfaces that separate them from underlying river and bayhead delta deposits. This was interpreted as evidence that rapid sea-level rise caused the observed landward shifts in these different coastal environments. A reduction in sediment supply to the bay could also have caused the landward retreat of bayhead deltas. However, this does not explain why the shoreline and associated tidal inlet appear to have shifted landward at the same time as the bayhead deltas. This style of coastal migration is referred to as *back-stepping*, interpreted as resulting from episodes of rapid sea-level rise. This interpretation was motivated by theoretical arguments for episodes of rapid grounding line retreat in Antarctica (Thomas and Bentley, 1978; Mercer, 1978).

Prompted by results from the offshore Trinity and Sabine River valleys, our first investigation of a modern estuary was conducted in Galveston Bay, where we hoped to obtain radiocarbon ages for some of the flooding surfaces observed in the offshore Trinity and Sabine River valleys. As with the offshore results, we found that the stratigraphic succession in Galveston Bay is composed of fluvial, bayhead delta, middle bay, and tidal delta deposits bounded by discrete flooding surfaces that record up-valley shifts in these depositional environments (fig. 6.9a, b). The up-valley shifts in bayhead deltas are most conspicuous and, when combined with radiocarbon ages, provide estimates of the timing and magnitude of Holocene flooding events. Seismic facies and sediment cores were used to construct paleogeographic maps depicting the changes caused by these flooding events (fig. 6.9c–h).

Initial flooding of the onshore portion of the Trinity Valley occurred ~9.6 ka and resulted in a long, narrow estuary with an extensive bay-

a

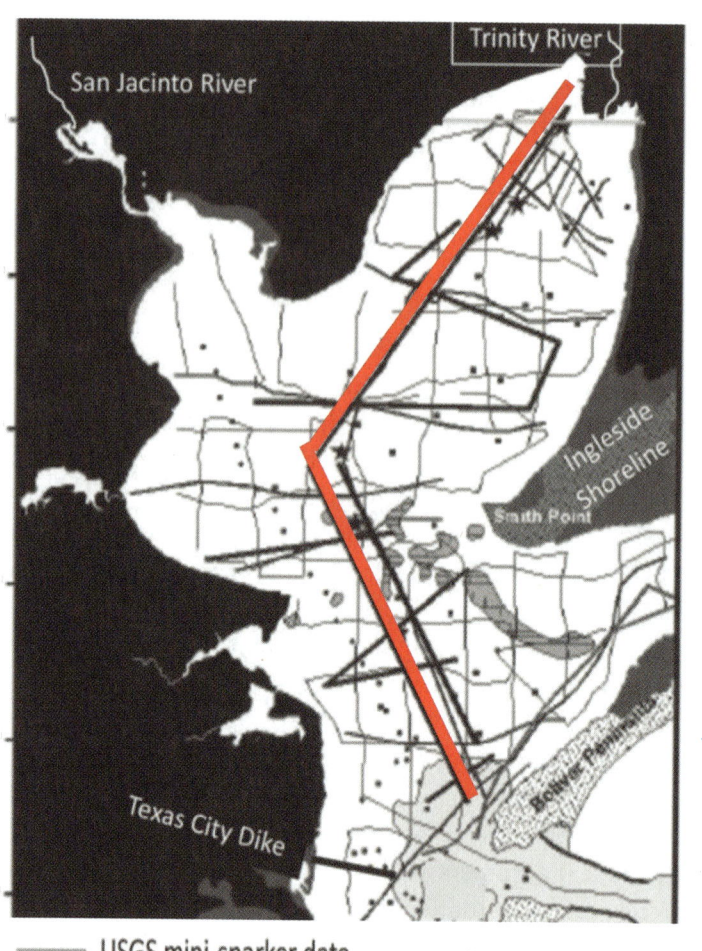

San Jacinto River

Trinity River

Ingleside Shoreline

Smith Point

Bolivar Peninsula

Texas City Dike

——— USGS mini-sparker data   ● Rice University vibracores
——— Rice University boomer data   ★ Rice University drill cores
——— Rice University CHIRP data

FIGURE 6.9. *Above and facing*, (a) Map showing seismic profiles and locations of drill cores used to study the evolution of Galveston Bay. (b) This north–south stratigraphic profile of Galveston Bay was compiled from a seismic profile and drill cores collected along the axis of the estuary. The red line on the map (a) shows the profile location. Four prominent flooding surfaces are recognized, and their ages are constrained using radiocarbon dates at ~9.6 ka, 9.1–8.9 ka, 8.2–8.0 ka, and 7.7–7.4 ka. The 7.7–7.4 ka flooding event resulted in an up-valley shift of more than 20 miles of the bayhead delta in approximately 400 years. Note also the late Holocene progradation of the Trinity bayhead delta. (c)–(h) Paleogeographic maps show magnitudes of change associated with flooding events. During the late Holocene, the major changes to Galveston Bay included flooding of the shallow margins of the estuary and changes in the Bolivar flood-tidal delta driven by the formation and lateral accretion of the Bolivar Peninsula. Modified from Anderson et al. (2008).

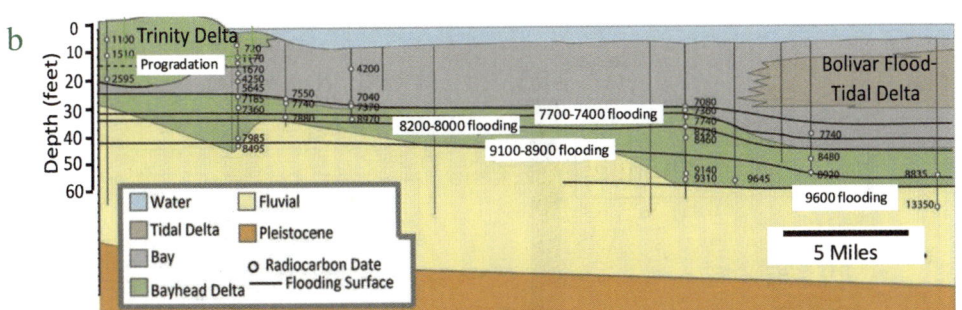

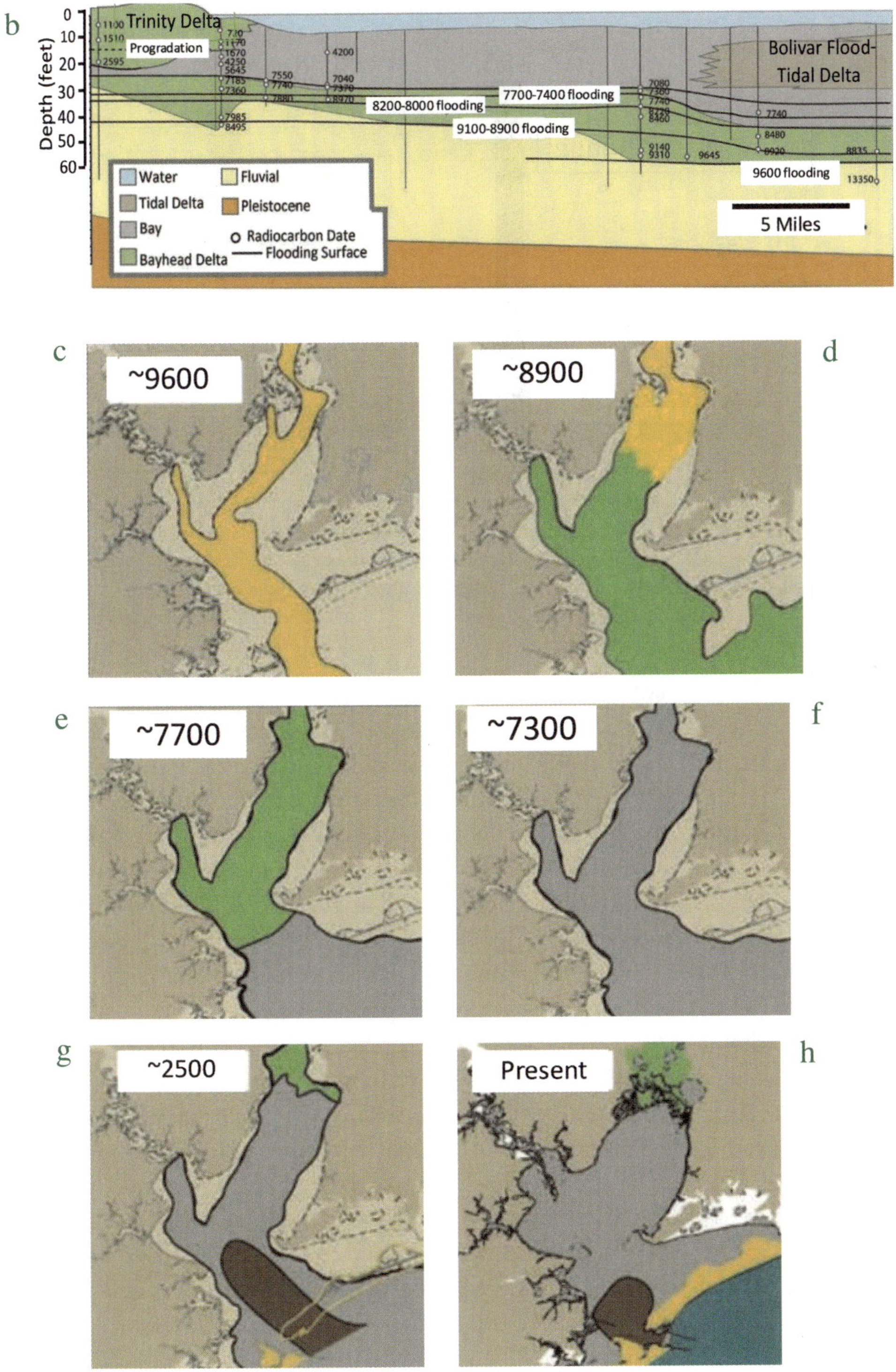

b

Depth (feet)

Trinity Delta
Progradation

Bolivar Flood-
Tidal Delta

8200-8000 flooding
9100-8900 flooding

7700-7400 flooding

9600 flooding

Water
Tidal Delta
Bay
Bayhead Delta

Fluvial
Pleistocene
○ Radiocarbon Date
— Flooding Surface

5 Miles

c   ~9600   ~8900   d

e   ~7700   ~7300   f

g   ~2500   Present   h

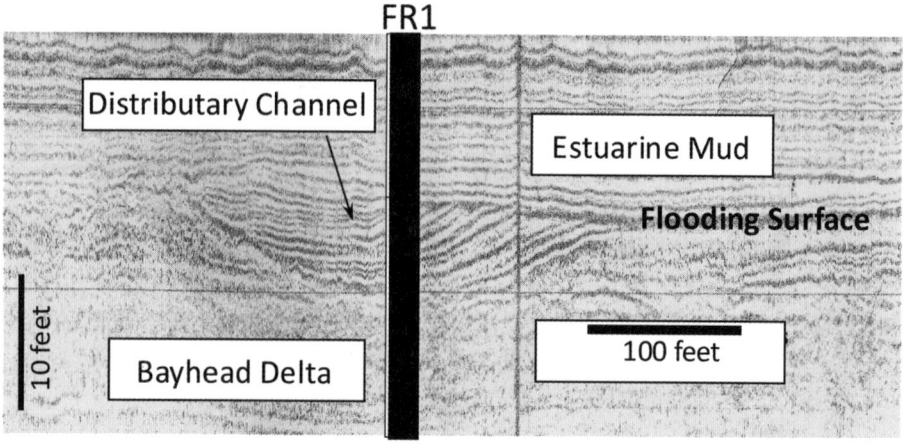

FIGURE 6.10. This seismic section shows bayhead delta deposits, including a sediment-filled distributary channel overlain by middle bay deposits. The surface separating these seismic facies is a flooding surface resulting in the delta plain's submergence. A drill core (FR1) collected near this seismic profile location supports this interpretation, and radiocarbon ages were used to establish the age of this flooding event at between 7.7 ka and 7.4 ka. Modified from Anderson et al. (2008).

head delta. Between ~8.9 ka and ~8.5 ka, the Trinity bayhead delta shifted landward across a relatively low-gradient portion of the river valley at an average rate of 0.7 mile per century. This surface merged up the valley with a younger, 8,200-year-old flooding surface whose magnitude was not determined because of an absence of clinoforms. During the flooding event ~7.7 ka to ~7.4 ka, the bayhead delta shifted up the valley about 20 miles in approximately 400 years.

The ~7.7 ka to ~7.4 ka event is one of the best-documented flooding events for the Gulf Coast and significantly impacted the estuary. The seismic image shown in figure 6.10 highlights the magnitude of the event. It shows a distributary channel, a subaerial feature, buried by estuarine mud. A sediment core (FR1) was used to verify these seismic facies interpretations. Radiocarbon ages from a nearby core sampled this same flooding surface and yielded ages of 7.7 ka and 7.4 ka below and above the surface, respectively.

Other investigated estuaries show a similar response to the sea-level rise during the early Holocene, characterized by landward shifts in estuarine environments. The magnitude of these flooding events varied with geographic location, which is expected given differences in the size and depth of the valleys occupied by these estuaries. But, given radiocarbon dating constraints and differences in the shapes of these valleys, it is clear that the same flooding events impacted multiple bays (Anderson et al., 2022). The broad geographic span of these bays and their different climate settings indicate that rapid sea-level rise was the most likely cause of observed flooding events. Given the rates and magnitudes of change during these flooding events, there is little doubt that they significantly impacted estuarine salinities and ecosystems. In Corpus Christi Bay, the middle bay area more than doubled, and the bayhead delta shifted landward approximately nine miles within a few centuries during the well-documented ~8.2 ka flooding event. At the same time, oyster reefs in the lower estuary were buried in mud, and new reefs were established in Nueces Bay (Simms et al., 2008; Goff et al., 2016). These changes were later correlated to a significant increase in salinity using fossil *dinoflagellates*, microfossils that are sensitive to salinity. Results from a core taken in the upper estuary showed a fivefold increase in dinoflagellates during the ~8.2 ka flooding event, which indicates an increase in salinity approaching open marine conditions (Ferguson et al., 2018a). This magnitude of change implies widespread submergence of Mustang Island, consistent with sediment cores through the island that sampled mostly intertidal deposits of this age (Simms et al., 2006). The estuary took nearly 300 years to recover from this salinity crisis.

The eastern Gulf Coast estuaries also changed significantly during the early Holocene, although only a few bays have been studied in detail. A study of Mobile Bay showed that during the ~8.2 ka flooding event, the valley was transformed from an extensive swampland to an open bay setting (Rodriguez et al., 2008). Tampa Bay and Apalachicola Bay also changed significantly during the early Holocene, including an episode of dramatic flooding in Apalachicola Bay between ~8.0 ka and 6.6 ka (Osterman et al., 2009), but details of the environmental changes in both estuaries are lacking.

The ~8.2 ka flooding event is now widely recognized as a global event (Rodriguez et al., 2010; Törnqvist and Hijma, 2012). The cause remains uncertain, but one potential water source was the rapid draining of glacial Lake Agassiz in north-central North America (Törnqvist and Hijma, 2012). In addition, the Antarctic Ice Sheet was retreating from the continental shelf during this time and was a likely contributor (e.g., Bentley et al., 2014; Prothro et al., 2020).

The magnitude of sea-level rise during this ~8.2 ka flooding event is poorly constrained, but estimates for the Gulf of Mexico range between 10

and 22 inches (Li et al., 2012). This is within the range of sea-level rise estimated for the collapse of the Thwaites Glacier in Antarctica. Even without
such an event, sea level is projected to rise about a foot by 2050 (fig. 1.2).

Case Studies

By ~7.5 ka, sea-level rise in the Gulf of Mexico had decreased from an average
rate of 6 inches/century to an average of 2 inches/century. Despite this
decrease in the rate of sea-level rise, estuaries of the Gulf Coast continued
to change. Most notable were shifts in the size and location of bayhead
deltas caused by variations in sediment delivery from rivers and changes
in tidal deltas that resulted from coastal barrier development.

Galveston Bay

Galveston Bay is a prime example of an estuary that continued to change
during the late Holocene. During this time, major flooding events ceased
as the wide, shallow margins of the Trinity Valley were slowly flooded,
and the Bolivar flood-tidal delta expanded (fig. 6.9c). During this same
period, the Trinity bayhead delta prograded approximately six miles as the
climate in east Texas was transitioning toward wetter conditions (Ferguson
et al., 2018b). Growth of the delta culminated with an increase in the rate
of progradation of nearly 2.5 miles during the past century. This more
recent growth phase is believed to have been caused by deforestation and
increased agricultural activity within the drainage basin (Phillips, 2010).
Growth of the delta has now ended, partly because of the construction of
the Trinity River dam. The result has been the formation of an extensive,
low-gradient delta plain mostly less than two feet in elevation.

The northern and western shorelines of Galveston Bay are home to the
largest urban and industrial complex on the Gulf Coast, so the estuary is
constantly threatened by chemical spills and urban runoff (fig. 6.11). Bottom
sediments in the Houston Ship Channel contain high levels of mercury,
lead, nickel, and zinc, as well as hydrocarbons and pesticides that accumulated mainly during the 1960s and 1970s (Dellapenna et al., 2020). These
pollutants remain within the upper layer of the sediment column where
resuspension, from either dredging or severe storms, remains a threat to
water quality.\<FIG 6.11>

The Houston Ship Channel and dredge spoil banks extend the entire
length of the estuary and have resulted in more focused freshwater discharge, including urban and stormwater drainage, along the bay's western

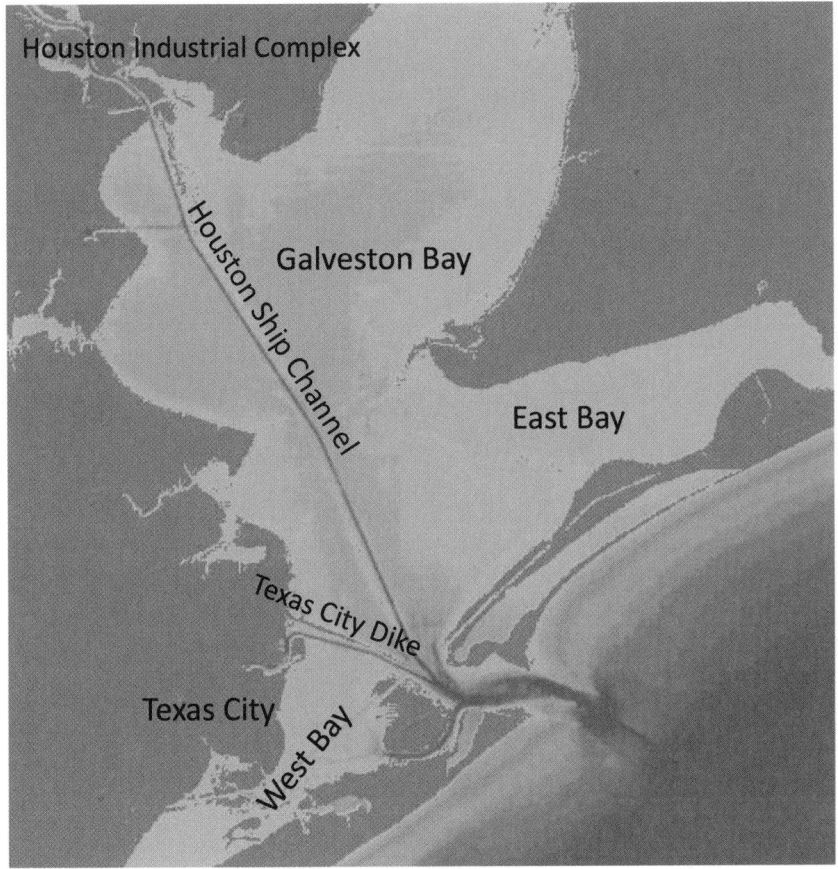

FIGURE 6.11. Bathymetric map of the Galveston Bay complex showing the Houston Ship Channel and the Texas City Dike, which have altered circulation within the estuary. From GeoMapApp.

side. Construction of the Texas City Dike, which extends across the juncture of Galveston Bay and West Bay, blocked the connection between these estuaries and resulted in higher salinities within West Bay, likely contributing to the demise of oyster reefs in the area. The next significant alteration of the Galveston Bay estuary could come with the proposed construction of a massive set of locks at the mouth of the estuary (chapter 9). The impact of these structures on circulation and salinity remains uncertain.

Figure 6.12 shows the latest NOAA coastal inundation map for Galveston Bay for the predicted one-foot sea-level rise by 2050. Based on this projection, the Trinity delta will shift approximately nine miles inland

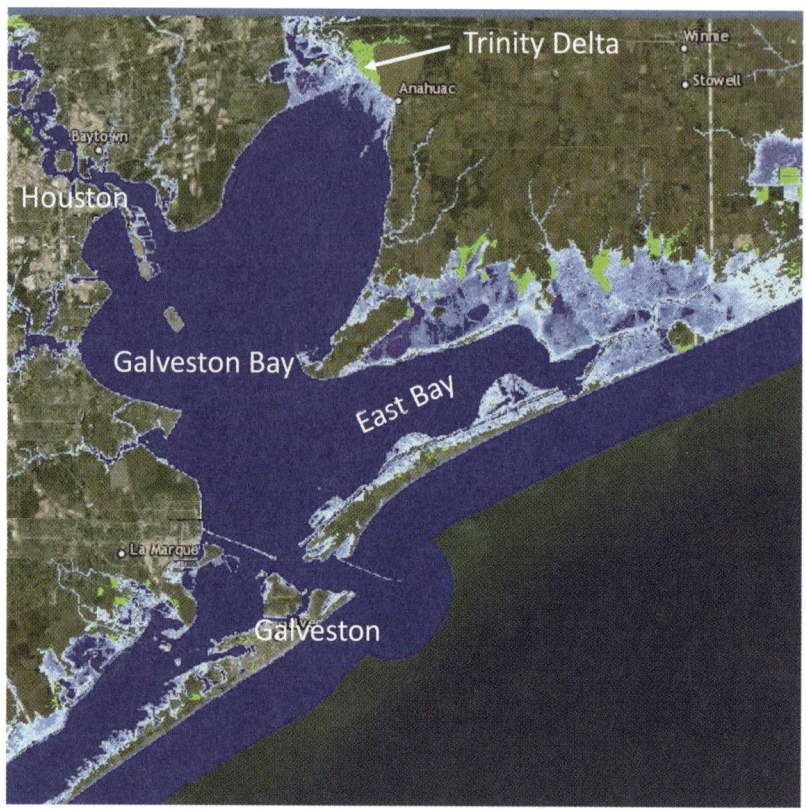

FIGURE 6.12. Model results from NOAA show changes to the Galveston Bay complex resulting from the one-foot sea-level rise projected for 2050, shown in light blue. Dark blue designates areas currently below sea level.

by this time. This is faster than the rate and magnitude of early Holocene flooding events, a result of late Holocene delta growth and the formation of an extensive, low-gradient delta plain. There will also be significant inundation north and east of East Bay, and the Bolivar Peninsula's fate remains uncertain. What is known is that these changes will mark a significant reversal from the growth of the delta and Bolivar Peninsula during the late Holocene.

Sabine Lake and Calcasieu Lake

Sabine Lake and Calcasieu Lake are among the more unusual Gulf Coast estuaries because they are connected to the Gulf of Mexico by narrow inlets that restrict tidal exchange with the Gulf (fig. 6.2f). This, coupled with the location of both estuaries within a humid climate zone, results in their relatively low salinities compared to estuaries to the west.

Before ~3 ka, both Sabine Lake and Calcasieu Lake were growing in size, and their bayhead deltas shifted northward as relative sea-level rise exceeded rates of delta plain accretion (fig. 6.13). Low sediment discharge

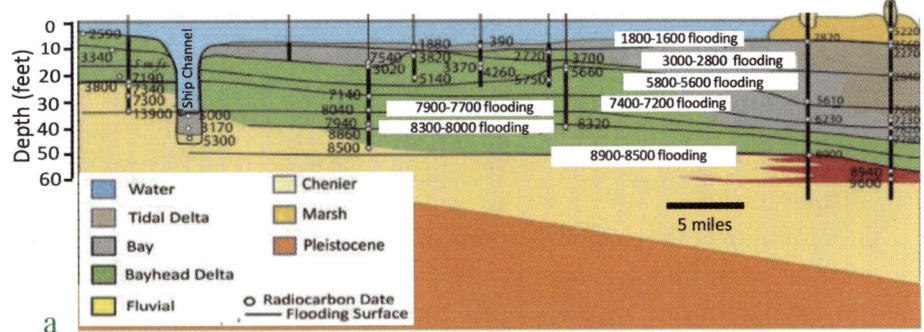

a

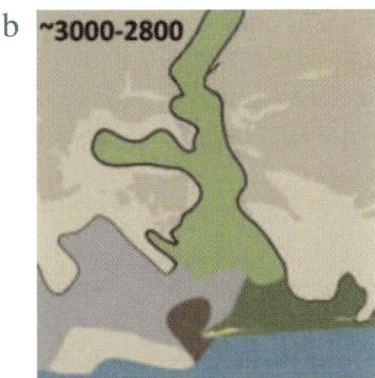

b ~3000-2800

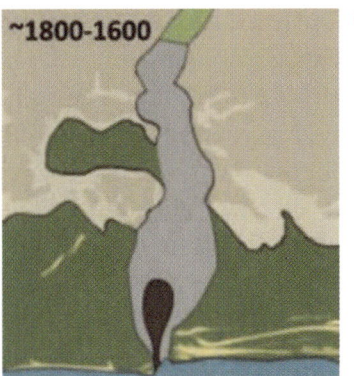

~1800-1600 c

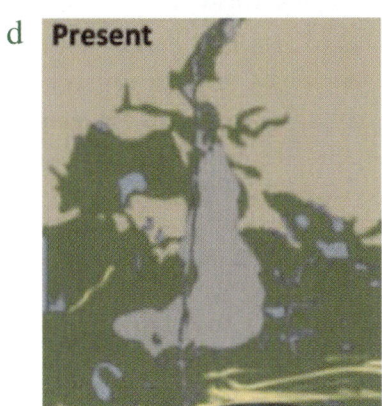

d Present

FIGURE 6.13. (a) A north–south stratigraphic profile for Calcasieu Lake was compiled from a seismic profile and drill cores collected along the axis of the estuary. (b)–(d) Paleogeographic maps illustrate the magnitude of change during the late Holocene. Modified from Milliken et al. (2008c).

to these estuaries is attributed to dense vegetation cover in the Sabine River and Calcasieu River drainage basins. Large flood-tidal deltas record wider tidal inlets, and higher salinities within both estuaries are indicated by more abundant oyster reefs and shells in older bay sediments (Milliken et al., 2008b, c). An increase in sediment supply from the Mississippi River resulted in chenier plain development and increased isolation of both estuaries as their tidal inlets became more restricted. By ~1.6 ka, the flood-tidal deltas in both estuaries no longer existed. Restricted tidal circulation resulted in greater sediment retention, and the shallow margins of both estuaries were filled with sediment and converted to wetlands. Meanwhile, the bayhead deltas in both estuaries continue to decrease in size and shift landward.

Infilling of the shallow margins of Calcasieu Lake and Sabine Lake to create extensive low-elevation wetlands was facilitated by human settlement in the region, marked by increased agricultural activity and extensive logging of east Texas pine forests. More recently, the trend in estuarine evolution has shifted to one of rapid drowning of these wetlands. Research by scientists at Tulane University has shown that 58 percent of the western Louisiana Chenier Plain is being inundated because of the acceleration of sea-level rise coupled with high subsidence rates. Some of the fastest rates occur in the low-lying areas between Calcasieu Lake and Sabine Lake. At the current rate of land loss in the region, Calcasieu Lake and Sabine Lake will be joined by 2050 (fig. 6.14).

Central and South Texas Coast

Estuaries of central and south Texas, including Matagorda Bay, San Antonio Bay, Copano Bay, Aransas Bay, Corpus Christi Bay, Baffin Bay, and Laguna Madre, experience semiarid to arid conditions and modest freshwater inflow from rivers and streams (fig. 6.1). These estuaries have limited connection to the Gulf of Mexico. As a result, they are characterized by relatively high salinities and hypersaline conditions in Baffin Bay and Laguna Madre.

Matagorda Bay is relatively large and has a unique shape that results from the drowning of a landscape with shore-parallel ridges (fig. 6.3). As with other Texas estuaries, its early Holocene evolution was dominated by sea-level rise and landward shifts in the Lavaca bayhead delta, which remained a prominent feature in the estuary during its early evolution (Maddox et al., 2008). By the late Holocene, freshwater inflow from the

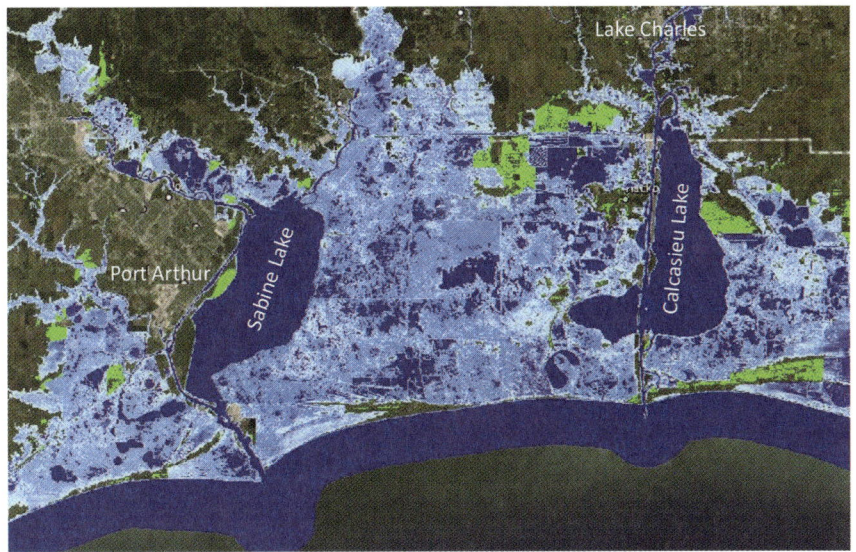

FIGURE 6.14. Model results from NOAA show portions of the west Louisiana and east Texas coast that will be drowned by the one-foot rise in sea level projected for 2050, shown in light blue. Dark blue designates areas currently below sea level.

Lavaca River could no longer sustain a large bayhead delta, but the estuary itself changed modestly. The upper bay, known as Lavaca Bay, is home to one of the largest Superfund sites in the United States, which includes Alcoa's now-inactive Point Comfort aluminum processing plant that dumped toxic wastes into the bay for decades. At one point, the Corps of Engineers proposed to dredge and expand the Matagorda Bay Ship Channel. This could have resulted in an environmental disaster by churning up mud contaminated with various toxic trace metals. Fortunately, environmental activists succeeded in halting the project, but a large hurricane could still resuspend bottom sediments.

Research in Corpus Christi Bay has provided a wealth of information about how it responded to sea-level rise and climate change during the Holocene (Simms et al., 2008; Rice et al., 2020). As with other Texas estuaries, the evolution of Corpus Christi Bay was dominated by rising sea levels during the early Holocene. The ~8.2 ka flooding event was especially impactful and resulted in the Nueces bayhead delta shifting landward ~8.0 miles in about two centuries (Simms et al., 2008). During the late Holocene, the Nueces bayhead delta continued to oscillate in size and location, including one episode of landward retreat at ~5.0 ka of up to 14 miles within a few centuries, and another significant flooding event at ~3.2 ka. There were intervening periods of delta progradation at ~3.8 ka and ~2.2 ka, with the delta prograding ~6 miles during its final growth phase. These dramatic changes in the estuary have been linked to alterations between warm/dry and cool/wet conditions and attest to its sensitivity to climate change (Rice et al., 2020).

A single tide-gauge record from Corpus Christi Bay shows an average historical RSLR rate of 21.3 inches/century, the second-highest measured rate in Texas. However, given the relatively high topography surrounding the estuary, it is expected to suffer a relatively minor impact from sea-level rise over the next several decades. Wave erosion has been a persistent problem, resulting in armoring of most of the estuary's western shoreline. The main impact of accelerated sea-level rise will occur in Nueces Bay, where NOAA predictions call for the bayhead delta plain to be inundated ~8.0 miles inland by 2050 (fig. 6.15). This is significantly faster than the ~8.2 ka inundation event and, as with Galveston Bay, is a result of aggradation of the delta plain since 2.2 ka.

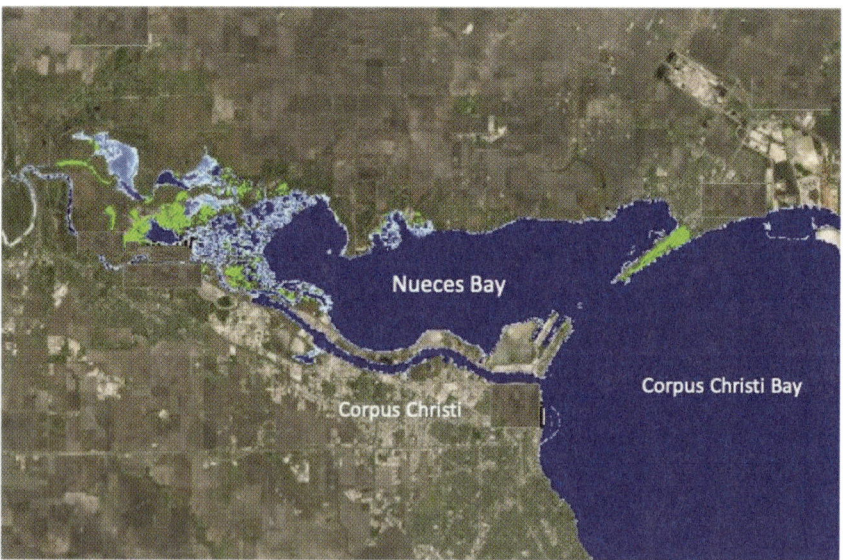

FIGURE 6.15. Model results from NOAA show the extent of the drowning of the Nueces bayhead delta that will result from the one-foot rise in sea level projected for 2050, shown in light blue. Dark blue designates areas currently below sea level.

The impacts of sea-level rise in the estuary are exacerbated by decreasing sediment delivery from the Nueces River, a result of increased aridity in the southwestern United States, and construction of the Wesley E. Seale Dam about six miles north of the delta. The dam was constructed in 1927 and reconstructed in 1955. More arid conditions are also expected to result in higher salinities for Corpus Christi Bay, resulting in a northward shift in oyster reefs and changes in vegetation patterns, including seagrass distribution.

Baffin Bay and Laguna Madre may well be the harbingers of future change for other estuaries of the central Texas coast. These estuaries are not connected directly to the Gulf of Mexico via a tidal inlet and receive limited freshwater drainage from streams. The salinity in Baffin Bay, as a result of its relatively small and highly variable freshwater input, restricted connectivity to the Gulf, and semiarid setting, varies between 2 and 85 parts per thousand (ppt), with an average salinity of 40 to 50 ppt. Hypersaline conditions in the estuary were established during the late Holocene as the region became increasingly more arid, and landward migration of Padre Island resulted in more restricted tidal exchange with the Gulf (Buzas-Stephens et al., 2014; Livsey and Simms, 2016).

Laguna Madre is a shore-parallel estuary that stretches some 130 miles from Corpus Christi Bay to the mouth of the Rio Grande (fig. 6.16). It is mostly less than three feet deep and is separated from the Gulf of Mexico by Padre Island. Connectivity to the Gulf is restricted to two relatively small navigation channels, one at the southern end of the lagoon and the other at Port Mansfield near the center of the lower estuary. This limited connection to the Gulf, in addition to little freshwater input from streams and an arid climate setting, makes it the only hypersaline lagoon in North America, with an average salinity of 45 ppt in the lower lagoon. The lagoon supports a rich and diverse ecosystem, which includes 80 percent of the state's coastal seagrass ecosystem.

The geological evolution of Laguna Madre is closely tied to that of Padre Island. Sediment cores from Laguna Madre that sampled lagoon and washover deposits dating back to ~5.4 ka indicate a larger, more open bay setting (Wallace and Anderson, 2010). For the last several millennia, South Padre Island was near its current location, and the lagoon began transitioning to its hypersaline condition. Today, South Padre Island has one of the fastest-retreating coastlines in Texas, with an average annual rate of nearly

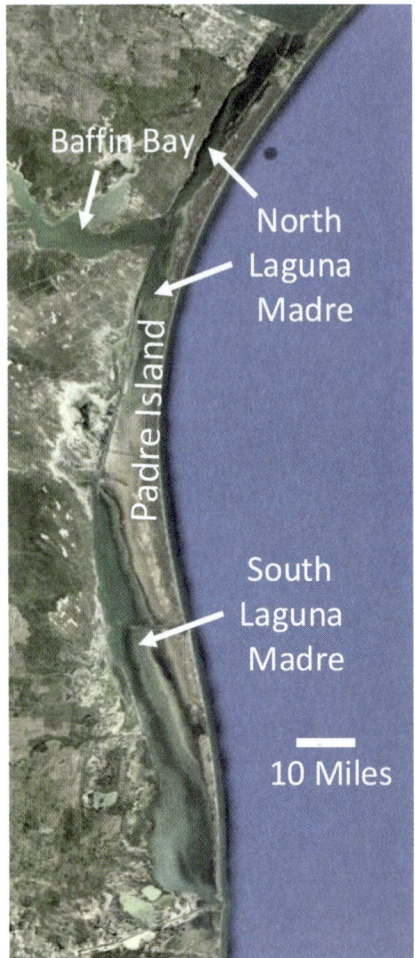

FIGURE 6.16. South Texas coast showing Padre Island and Laguna Madre. From Google Earth.

−6 feet/year and rates as high as −18 feet/year south of Port Mansfield. As the island shifts landward, it loses elevation and narrows, making it more vulnerable to storm breaching and washover. Sediment cores from the lagoon show increased rates of washover through time. This vulnerability is expected to increase with accelerated sea-level rise. More arid conditions are expected to increase salinities within the estuary.

Louisiana Estuaries

Larger Louisiana estuaries include Chandeleur Sound, Barataria Bay, Terrebonne Bay, Atchafalaya Bay, and Vermilion Bay (fig. 6.1). The evolution of these estuaries is closely linked to that of delta lobes and barrier

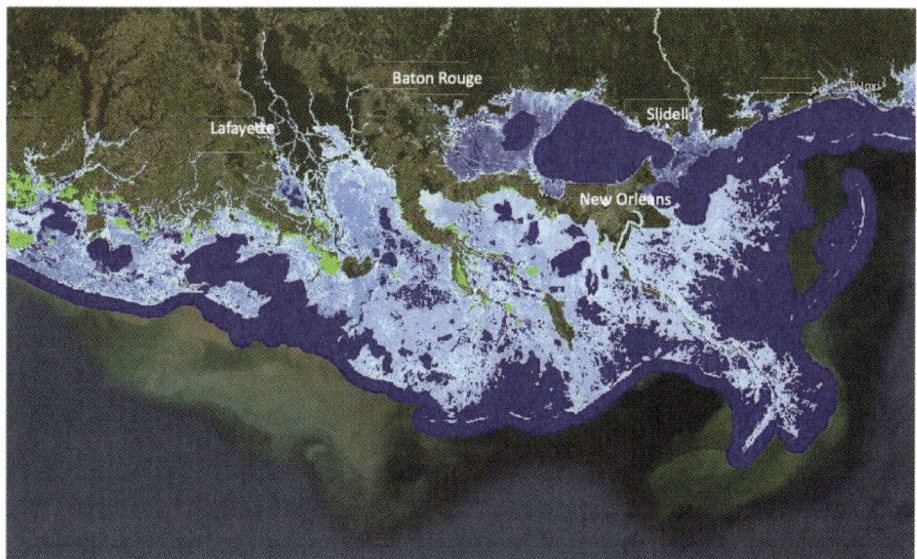

FIGURE 6.17. Model results from NOAA show portions of south Louisiana that will be inundated by the one-foot rise in sea level projected for 2050, shown in light blue. Dark blue designates areas currently below sea level.

islands formed during the late Holocene in response to the shifting and abandonment of the Mississippi River Delta lobe, the timing of which varied across the region (fig. 4.1). They are highly productive ecosystems and the heart of Louisiana's seafood industry. All will experience significant change by 2050 as the south Louisiana coast is inundated by rising sea levels (fig. 6.17). The main result will be an expansion of shallow estuaries and a reduction in the extent of wetlands. This change is occurring so fast that its ecosystem impacts remain uncertain. Another impact will be a reduction in the size and extent of barrier islands. These combined effects will result in altered salinities and increased vulnerability to storm impacts.

Mississippi Sound

With a surface area of nearly 3,000 square miles, Mississippi Sound is the largest estuary on the Gulf Coast. It extends about 75 miles from Mobile Bay to Louisiana and averages 10 miles from the mainland to a series of barrier islands separated by numerous broad tidal inlets (fig. 6.2d). It includes several smaller coastal estuaries, including St. Louis Bay, Biloxi Bay, Pascagoula Bay, and Grand Bay. Freshwater drainage into the sound comes from the Escatawpa, Pascagoula, Tchoutacabouffa, Biloxi, Wolf, and Jourdan Rivers and numerous streams. The large area of the sound and the small barrier islands separated by wide inlets promote relatively high salinities (average 24 ppt). Given its large size, winds play a greater role in sound circulation than in other Gulf Coast estuaries.

Radiocarbon ages from sediment cores collected within Mississippi Sound indicate that initial flooding occurred ~9.0 ka and that the barrier island chain began to form ~6.0 ka (Otvos, 1979; Hollis et al., 2019). Sediment cores from the sound record only modest change during the late Holocene, despite rapid lateral growth of barrier islands toward the west between ~4.0 ka and ~2.5 ka. Historically, the offshore barrier islands have experienced significant breaching and changes in size, shape, and location (fig. 5.23). But given the large extent of tidal inlets between the islands, there has been little permanent impact on the estuary. In Alabama, the northern shoreline of the sound has remained relatively pristine and covered by extensive wetlands. In Mississippi, the coast has been significantly developed, resulting in the loss of approximately 10,000 acres of coastal wetlands.

The greatest threat to Mississippi Sound comes from humans. One of the most significant impacts came from the Bonnet Carré Spillway. It was constructed as a flood control structure to divert floodwaters from the Mississippi River through Lake Pontchartrain and Lake Borgne and into Chandeleur Sound and Mississippi Sound. The main impact has been extended periods of significantly reduced salinities in the sound following floods and increased drainage from the spillway. At times, salinities in the western sound have decreased significantly for two to three months. This has resulted in significant impacts on the region's seafood industry, in particular the shellfish industry. Other significant impacts on the sound include hurricanes, particularly Hurricane Katrina, which devastated this area of the coast, and, more recently, the Deepwater Horizon oil spill.

Mobile Bay

The elongated shape of Mobile Bay was inherited from the flooding of the Mobile-Tensaw incised valley (fig. 5.24). It is one of the deepest river valleys on the Gulf Coast, a reflection of it being one of the largest drainage systems in the contiguous United States, with a combined drainage basin area of 84,195 square miles. This, combined with its humid climate setting and relatively high relief within the drainage basin, has resulted in a long history of high sediment delivery to the coast and the largest bayhead delta on the Gulf Coast. In his book *Saving America's Amazon*, author Ben Raines refers to the Mobile bayhead delta as the most biodiverse ecosystem in our nation.

The evolution of Mobile Bay was the focus of research by the USGS and Alabama Geological Survey in the 1990s, followed a decade later by a collaborative research program between Rice University and the University of Alabama (Rodriguez et al., 2008). The latter study was built on seismic data acquired during the earlier USGS studies and included acquiring additional seismic data and several drill cores within the estuary. The analysis of the drill cores showed that between ~8.6 ka and ~8.2 ka the Mobile incised valley was flooded to create a vast swampland. At ~8.2 ka, a significant flooding episode resulted in rapid inundation of the low-gradient valley floor to form an open bay environment. Since then, there has been little change in the estuary, including its bayhead delta. This long-term stabilization of the bayhead delta indicates relatively little change in sediment discharge to the estuary. This further implies that the drainage basins of rivers flowing into the bay experienced little significant climate change during the late Holocene. However, the evolution of the Morgan Peninsula and Dauphin Island caused substantial changes to the estuary. By ~5.3 ka, the eastern portion of Dauphin Island had formed, and the Morgan Peninsula had accreted to near its current location, constraining the mouth of the estuary and leading to the formation of the Mobile tidal delta (fig. 5.5a).

Although the Mobile bayhead delta grew throughout the late Holocene, this trend has changed in historical time and the delta has begun to retreat inland. The estuary lost over 10,000 acres of emergent wetlands between the 1940s and 1979 (Duke and Kruczynski, 1992). Most of this loss is attributed to industrial, navigational, and urban development and the construction of multiple dams. Deforestation has partially offset these impacts, which has increased sediment supply to the delta. One of the most significant impacts on the delta was the construction of the Mobile Bay causeway in the late 1920s, which altered sediment delivery to the upper bay. Construction of the causeway also contributed to the blockage of nutrient-rich waters and the encroachment of invasive aquatic plant species that threaten the delta's delicate ecosystem. The Mobile River has been ranked as the third most polluted river in the United States, owing mainly to high trace metal concentrations in bay sediments. The trace metals were delivered to the river and bay from coal ash in multiple ash storage sites and from a large aluminum processing facility that drained directly into the river for decades. As with Galveston Bay and Lavaca Bay, resuspension of these contaminated sediments by dredging or by a hurricane poses a significant threat to Mobile Bay's water quality and ecosystem.

Excluding the Mississippi River Delta, the Mobile bayhead delta has the largest low-elevation (<2 feet) delta plain on the Gulf Coast. A one-foot sea-level rise will result in ~25 miles of landward migration of the front of the delta by 2050. This is equal to a loss of nearly 310 square miles of wetlands (fig. 6.18). So, even though Mobile Bay has remained relatively unchanged during the past few thousand years, it is expected to suffer significant change in the next few decades as the Mobile bayhead delta shifts landward. The magnitude of change is expected to exceed changes at any other time during the Holocene, including the ~8.2 ka flooding event.

Apalachicola Bay

Apalachicola Bay is a relatively shallow estuary that is approximately 40 miles long and shielded from the Gulf of Mexico by St. George, Little St. George, and St. Vincent Islands, with a narrow inlet between Little St. George and St. Vincent Islands (fig. 6.2a). It is located at the mouth of the Apalachicola River, which receives most of its drainage from the Chattahoochee and Flint Rivers. Historically, the mixture of fresh water and Gulf waters in the estuary has been optimal for various fish, crabs, and Florida's largest oyster population, making it the state's most prolific estuary.

A study of Apalachicola Bay by the USGS in 2005 and 2006 aimed to better understand the estuary's evolution. Just three of the cores collected during the survey sampled the early Holocene section, so little is known about the estuary's past response to rapid sea-level rise. The record shows that by 6.4 ka, the estuary had formed behind shallow offshore sandy shoals and spits. By ~4.4 ka, these shoals had become barrier islands surrounding the estuary, with estuarine salinities suitable for the growth of oyster reefs.

Apalachicola Bay is experiencing a reduction in freshwater inflow from the Chattahoochee and Flint Rivers caused by agricultural and urban usage. Much of this usage comes from Georgia, including the greater Atlanta region. The resulting changes in the estuary's salinity regime and decades of overharvesting have led to a decline in Apalachicola's oyster industry and prompted an ongoing battle between Florida and Georgia over water rights. But the greatest threat to the estuary is rising sea level, which, at the current rate of rise and with limited upstream topographic constraints, will virtually inundate the current bayhead delta by 2050 (fig. 6.19).

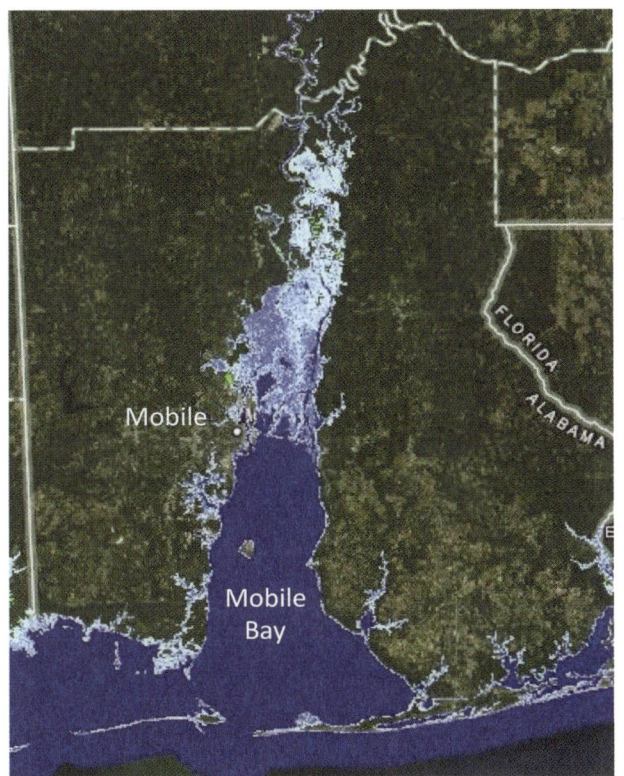

FIGURE 6.18. *Left*, Model results from NOAA show portions of the Mobile Bay coastal setting that will be inundated by the one-foot rise in sea level projected for 2050, shown in light blue. Dark blue designates areas currently below sea level.

FIGURE 6.19. *Below*, Model results from NOAA show portions of Apalachicola Bay that will be drowned by the one-foot rise in sea level projected for 2050, shown in light blue. Dark blue designates areas currently below sea level.

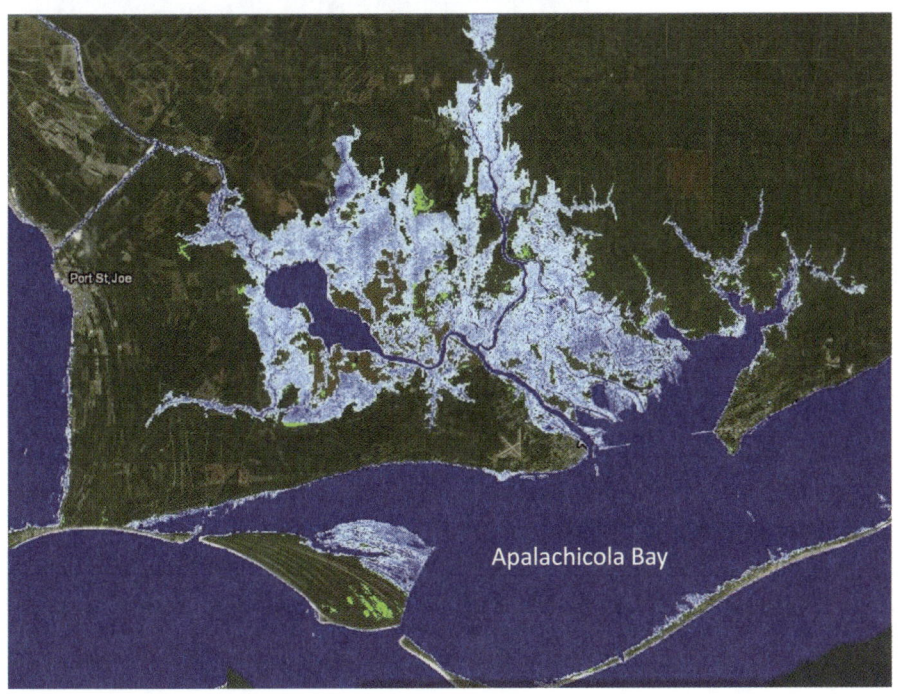

Tampa Bay

Florida's southwest coast includes numerous shore-parallel bays but
only two large river-influenced estuaries, Tampa Bay and the Gasparilla
Sound–Charlotte Harbor estuary (fig. 6.1). This is because there are only
two prominent river valleys in the region, associated with the Caloosahatchee River and Peace River drainage systems.

Tampa Bay has a highly irregular shape and bathymetry that results
from river valley incision and subsequent erosion of the estuary's limestone
substrate. No bayhead delta exists because of the limited sediment supply
from rivers and streams. Most of the estuary's sediment fill is composed of
bay mud that accumulated slowly during the past ~7,000 years. As a result,
there is little stratigraphic record of the estuary's past response to Holocene
sea-level rise and climate change.

Given the relatively high coastal relief of the Tampa Bay shoreline and
the absence of a bayhead delta, it will not be as strongly impacted by sea-
level rise as most other Gulf Coast estuaries. The greatest threat is from
urbanization in the areas surrounding the estuary. This includes Tampa, St.
Petersburg, Clearwater, and Bradenton, which grew and expanded rapidly,
mainly after the 1950s. During this period of growth, extensive portions of
coastal wetlands surrounding Tampa Bay, Boca Ciega Bay, Old Tampa Bay,
and Hillsborough Bay were filled for residential and commercial development. Currently, the bay is rimmed with freshwater drainage outlets from
these urban areas, which has resulted in increased nutrient influx. As a
result, the estuary is filled with organic-rich mud, referred to as "muck" by
the locals. Given the relatively shallow depth of the estuary, this sediment
is subject to resuspension during hurricanes, which could result in a water
quality disaster. Even without such an event, the complex circulation in the
estuary poses a challenge in predicting the dispersal and fate of pollutants
in Tampa Bay (fig. 6.5).

Geological studies have revealed rapid flooding events that drastically
altered river-influenced estuaries across the Gulf Coast during the early
Holocene, marked by landward shifts in environments at rates as high as 6
miles/century. Sea-level records indicate that these events occurred when
relative sea-level rise was within the range (~20 to ~40 inches/century) at

which vertical accretion rates in wetlands are exceeded by sea-level rise. Given current and predicted rates of sea-level rise, relatively high subsidence rates caused by compaction of thick Holocene sediments within these estuaries, and reductions in sediment supply, the bayhead deltas within larger estuaries are expected to be significantly inundated by 2050. Indeed, the magnitude of inundation will exceed flooding events during the early Holocene. This is mostly because the slow sea-level rise of the late Holocene resulted in the aggradation of bayhead deltas and the formation of extensive low-gradient delta plains and adjacent floodplains. These areas will experience widespread inundation by midcentury, including bayhead deltas in Nueces Bay, Galveston Bay, Sabine Lake, Calcasieu Lake, Mobile Bay, and Apalachicola Bay. Based on geological records, the ecosystem impacts on these estuaries will be significant. However, the nature and magnitude of these ecosystem changes remain problematic. Research is needed to investigate how this rapid flooding of bayhead deltas will impact estuarine ecosystems and, ultimately, the Gulf of Mexico ecosystem.

Significant changes continued in river-influenced estuaries of western Louisiana and Texas during the late Holocene, when the rate of sea-level rise averaged only 2 inches/century. The timing and magnitude of these events varied across the region, indicating that they were caused by climate change and associated changes in sediment supply from rivers and streams. In south Texas, increasing aridity during the late Holocene caused a shift to higher estuarine salinities. A similar shift is expected this century and may impact estuaries farther north. The changes that are occurring today are exacerbated by human influence, which includes alteration of tidal inlets and estuary bathymetry associated with navigation channels, alteration of freshwater discharge caused by dams, direct alteration of river and stream channel systems, and dramatically altered industrial and urban runoff into these water bodies. While the dumping of contaminants into estuaries has been curtailed in recent decades, these pollutants have not gone away. Several estuaries contain sediments with highly concentrated organic and inorganic contaminants from decades of industrial and urban runoff. Resuspension of these sediments during hurricanes threatens the estuarine ecosystem and human health. The recent near miss of Tampa Bay by Hurricanes Ian, Idalia, Helene, and Milton should serve as a reminder of this threat.

When I was 12, my father gave up on his dream of being a party boat fisherman, so we sold our house on the Morgan Peninsula and moved to Satsuma, Alabama, just a few miles west of the Mobile Bay delta. I spent my teenage years exploring the delta and appreciating its beauty and uniqueness. As my friends and I traveled down the Mobile River, we would journey a few miles south from pristine cypress swamps filled with alligators and other wildlife to a landscape lined with shipyards, paper mills, chemical plants, and substantial tailing piles of an Alcoa processing facility whose red dust spread across the landscape. My father and his fishing buddies would catch mud minnows in the small streams flowing from the tailings into the Mobile River. They would joke that those minnows were excellent bait because they had survived that polluted water. The visibly deformed minnows were supposedly the best candidates for luring a large flounder.

After I moved to Houston, I would return to visit my parents, who spent the remainder of their lives in Satsuma. Those trips provided an opportunity to watch the painfully slow efforts to clean up the Mobile River, a process that drags on even now. However, the Mobile delta faces a new threat far more devastating than the polluted waters from paper mills and chemical plants. The latest NOAA predictions show that by the time my grandchildren are in their 50s, Satsuma will be bordered to the east by a shallow bay, and there will be cypress graveyards where the lush delta plain once existed. It is hard for me to reconcile that humans will have inflicted more change on this magnificent delta in just three decades than has occurred in thousands of years.

7

Demise of Wetlands and Seagrass Meadows

After decades of onslaught from urban and industrial development and river flow alteration, wetlands and seagrass meadows are now faced with the impacts of accelerating sea-level rise. Most Gulf Coast wetlands occur at elevations of less than two feet, and natural and artificial barriers block migration to higher ground. Survival hinges on their ability to accumulate sediment and maintain upward growth fast enough to outpace sea-level rise, a battle whose outcome is determined at a scale of a fraction of an inch per year. Most Gulf Coast wetlands are losing this battle and will

vanish by 2050. Likewise, seagrass meadows are starting to vanish at alarming rates, and the Gulf of Mexico ecosystem will be significantly impacted by their loss.

The Gulf Coast is home to over half of the nation's coastal wetlands. Seagrass meadows occur below low tide level in locations where relatively clear waters exist, such as Laguna Madre, Texas, and along portions of the Mississippi, Alabama, and Florida coasts. These wetlands and seagrass meadows are spawning grounds for crustaceans, mollusks, and small fish that are essential to the Gulf Coast ecosystem and permanent and seasonal habitat for a majority of North America's bird species. In addition to these ecosystem services, wetlands play a vital role in the carbon cycle by sequestering significant amounts of CO_2 and helping to filter water flowing from rivers and streams into estuaries and the open Gulf. They are also crucial in absorbing wave energy and reducing storm tides during hurricanes. Given their enormous value, it is especially alarming that the loss of Gulf Coast wetlands far exceeds losses anywhere else in the United States. Louisiana, Florida, and Texas are among the five states expected to see the most extensive wetland loss this century. This includes the rapid loss of bald cypress wetlands, which have flourished throughout the Holocene.

Gulf Coast wetlands are distinguished by their low-diversity plant communities, which change in an onshore direction with increasing elevation (fig. 7.1). Nearest the coast, saltmarsh vegetation is dominated by *Spartina alterniflora*, which is adapted to frequent tidal inundation. Saltmarshes transition landward into brackish marshes, where salinities vary widely because of infrequent marine inundation. Their vegetation is dominated by *Juncus roemerianus* and *Spartina patens*. Farther inland, freshwater (cattail) marshes provide crucial habitats and drinking water for waterfowl and other fauna. The south Florida coast, including the Everglades, is occupied by mangrove wetlands.

The geological record of the Gulf Coast indicates that the late Holocene was a period of growth and expansion of estuaries and deltas and their associated wetlands and seagrass meadows. This long-term pattern ended by the mid-1800s, when humans assumed a crucial role in the destruction of wetlands. The Swamp Land Acts of 1849, 1850, and 1860 were aimed at destroying wetlands by declaring them useless land that should be drained and converted to agricultural land. It is estimated that these acts paved the way for roughly half of the nation's wetlands to be converted to agricultural

a

b

FIGURE 7.1. (a)–(d) Generalized vegetation zones for Gulf Coast wetlands.

c

d

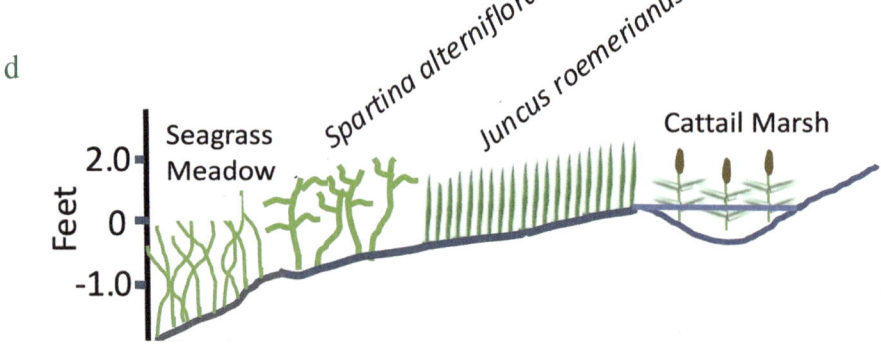

FIGURE 7.1. *continued*

land. This practice continued until the 1970s when broader recognition of the value of wetlands as wildlife habitats and nursing grounds for many marine organisms led to legislation aimed at their protection. The passage of the Clean Water Act in 1972 was a significant achievement in wetland conservation. It prohibited the release of pollutants into water bodies and protected the overall quality of surface waters. However, poor compliance and enforcement related to industrial and urban development continue today because new regulations and their enforcement were left to individual state agencies, with mixed results. Some Gulf Coast states are among the worst offenders. Meanwhile, rapid growth and development of numerous coastal counties continue in the Gulf Coast region, under increasing threat from rising sea levels, alterations of water and sediment supply routes to the coast, salinity changes caused by alteration of coastal hydrology, and direct impacts from hurricanes and oil and chemical spills.

The National Oceanic and Atmospheric Administration (NOAA) and the US Fish and Wildlife Service are actively engaged in measuring wetland loss and its causes. NOAA's Coastal Change Analysis Program (C-CAP) estimated wetland losses in the United States between 1996 and 2006 to be approximately 400 square miles. More recent progress in wetland monitoring has been driven by new satellite technology and more sophisticated remote-sensing methods that provide the spatial resolution needed for detailed mapping of coastal elevation and wetland types. NOAA's Coastal Inundation Dashboard provides detailed projections of coastal change caused by sea-level rise in the coming decades and centuries. The dashboard shows alarming rates of coastal change by 2050, with most of that change being wetland inundation (fig. 1.2). The US Fish and Wildlife Service has developed an interactive National Wetlands Inventory website that provides maps of coastal wetlands, distinguished as estuarine, marine, and freshwater wetlands (https://www.fws.gov/program/national-wetlands-inventory). This improved mapping provides the detail needed to address rates and causes of wetland loss and will play a key role in monitoring future changes.

Forecasting Wetland Loss

Predicting wetland loss in response to sea-level rise and climate change is challenging. Early predictions were based on simple "bathtub models" where coastal plain topography was flooded without considering sediment supply and other factors. These models yielded eye-opening results but failed to capture the complexities of wetland inundation over the past several decades. In the mid-1980s, a group of scientists in Massachusetts developed the Sea Level Affecting Marshes Model (SLAMM), which attempts to simulate processes involved in wetland inundation during long-term sea-level rise. It incorporates sedimentation and accretion rates and provides output that can be viewed with Geographic Information System (GIS) software, including relative sea-level change at 5- to 25-year increments. Unfortunately, the SLAMM results are only as good as the data input to the model, and historical observations clearly show that wetland loss is complicated by regional variations in subsidence, terrigenous sediment supply from rivers and streams, rates of storm washover, and wave erosion. The relative importance of these different factors in wetland loss varies across the Gulf Coast. In Louisiana, subsidence and the supply and dispersal of sediment are the most critical variables. Studies in Texas during the 1970s and 1980s showed the highest rates of wetland loss in areas with the greatest exposure to winds and waves. Now, sea-level rise poses the greatest threat to the state's wetlands. NOAA's Coastal Inundation Dashboard provides the best predictions for wetland loss in the Gulf Coast region, but it is limited by a paucity of the data needed to reconcile the causes. Field measurements of subsidence, sediment supply and dispersal, vertical accretion rates, and other relevant data are needed to fill that gap.

Response to Sea-Level Rise

Vertical accretion is the upward growth rate of accumulating new sediment (both terrigenous and organic), and it is an essential influence on wetlands' capacity to combat sea-level rise. When the vertical accretion rate is below the RSLR rate, the wetlands drown to become open water bodies (Morris et al., 2002). Studies have shown that wetland inundation occurs when RSLR exceeds ~0.4 inch/yr (10 mm/yr) (Kirwan et al., 2016). This is consistent with results from geological studies of Gulf Coast wetlands showing widespread inundation when rates exceed this value (Törnqvist et al., 2021; Anderson et al., 2022). This includes late Holocene flooding events caused by decreases in sediment supply. So, a small increase in the rate of sea-level rise or a decrease in sediment supply can tip the scales in favor of wetland inundation.

Results from a global-scale study of coastal wetland extent led to the argument that the area of coastal wetlands could increase this century (Schuerch et al., 2018). However, this scenario includes the provision of room for landward migration. Results from more recent work showed that the area available for landward migration of coastal wetlands is considerably less than the current wetland area (Osland et al., 2022). This is true for most of the Gulf Coast, where the landscape of the coastal plain includes abrupt steps in elevation associated with Pleistocene shorelines and fluvial terraces, as well as artificial flood control levees and dredge spoil banks lining the Intracoastal Waterway (fig. 6.3). It has also been argued that elevated CO_2 levels may enhance vertical accretion and counterbalance sea-level rise (Kirwan et al., 2016). However, research has failed to support this argument (Zhu et al., 2022).

Severe storms have also taken their toll on wetlands and seagrass meadows. While this is a natural process and these environments are well suited for recovering from storms, there is concern that larger and more powerful storms leave a more lasting impact and may play a more significant role in the loss of wetlands and seagrass meadows. An example is the back-to-back hurricanes Katrina and Rita in 2005, which had devastating and lasting effects on coastal marshes stretching from east Texas to Mississippi. The wetlands of Florida's Big Bend area and the Everglades occupy thin organic mud resting on limestone. Storm erosion of this thin mud layer could result in widespread wetland loss. Oil spills have also affected Gulf Coast wetlands, but these impacts are generally less severe in their geographic range and duration. The exception was the 2010 Deepwater Horizon spill, which released almost five million barrels of oil into the marine and coastal environment, damaging coastal habitats along over 1,000 miles of the Gulf Coast.

South Louisiana's Vanishing Wetlands

Regarding wetland loss, Louisiana is the Gulf Coast "poster child." Dredging and alteration of the Mississippi River and its main tributary, the Missouri River, during the twentieth century led to a significant reduction in sediment supply to south Louisiana wetlands. The result was a ~25 percent reduction in the area of wetlands (~2,000 square miles) between 1932 and 2016 (Couvillion et al., 2017). This is the world's highest rate of wetland loss over the past century. Sadly, it pales in comparison to the predicted loss for the next 30 years by Louisiana's Coastal Master Plan and NOAA's Coastal Inundation Dashboard (figs. 6.14 and 6.17). This bleak prediction is supported by a recent analysis by scientists from Tulane University, which led to the alarming conclusion that the loss of Louisiana's remaining 930 square miles of marshland is inevitable (Törnqvist et al., 2020). Based on measurements from 253 sites throughout coastal Louisiana, vertical accretion rates at about 90 percent of these sites are not keeping pace with the current RSLR (Li et al., 2024).

The argument has been made that land loss in south Louisiana is the result of natural causes, particularly the high rates of subsidence characteristic of the world's large deltas. However, deltas are by nature areas of growth, or at least that was the case before modern times and the unprecedented loss of land in the great deltas of the world (Syvitski et al., 2009). The Mississippi Delta is no exception to this trend. Its growth over the past several thousand years is well documented (fig. 4.1), but so is its rapid demise (fig. 6.17). It has become the fastest-disappearing delta, with wetlands accounting for most of that land loss. The reversal in the delta's development is widely accepted by the scientific community as being caused by artificial hydrological modifications, including alteration of the river's natural drainage system and construction of flood control structures, amplified by increased rates of RSLR (Day et al., 2007).

Wetland loss in Louisiana is not restricted to the Mississippi Delta. It includes the chenier plains of western Louisiana, where the natural drainage system has been less altered by humans. Scientists estimate that only 42 percent of the wetlands in the chenier plain can keep pace with the present-day rise in sea level (Jankowski et al., 2017). This is because relatively high subsidence rates characterize the chenier plain, but it receives less terrigenous sediment input than the Mississippi Delta. That percentage is decreasing as the rate of sea-level rise increases. So, most of Louisiana's wetlands will be inundated by 2050. Most of the flooded area will be converted to shallow bays that will be less effective in dampening storm surges. More concerning is the changing role of these shallow water bodies in the delta's nutrient production, carbon sequestration, and ecosystem. The region of wetland loss extends into east Texas, including the lower Sabine River valley (fig. 6.14) and the vast stretch of wetlands extending westward to Galveston Bay (fig. 6.12).

Wetland Loss in Texas

Most Texas wetlands occur along the fringes of estuaries or occupy bayhead deltas. Slow sea-level rise during the late Holocene provided ideal conditions for their growth, resulting in extensive low-gradient delta plains. That long-term pattern ceased during the past century, and erosion and inundation are now gnawing away at the state's wetlands.

Estimates of vertical accretion in bayhead deltas are higher than those measured in wetlands that fringe estuaries because of the high terrigenous sediment supply to deltas. This explains why wetlands associated with bayhead deltas grew significantly during the late Holocene. The Trinity bayhead delta in Galveston Bay prograded nearly five miles between 2.6 ka and 1.6 ka as the climate in east Texas was transitioning from more arid to more humid conditions (Ferguson et al., 2018b). During the same interval, the Nueces bayhead delta experienced fluctuations in growth and retreat that have been linked to warm/dry and cool/wet climate oscillations (Rice et al., 2020). These observations convey that climate is an essential player in wetland development. The role of climate is especially important in south Texas, where climate models predict more arid conditions (Nielsen-Gammon et al., 2020). The result will be decreased sediment supply from rivers and changing conditions for wetland vegetation, which, coupled with the acceleration in sea-level rise, will result in significant wetland loss during the next several decades. Based on the NOAA dashboard, roughly 530 square miles of the state's wetlands will be inundated by 2050. Unfortunately, there is a paucity of the data needed for a more accurate estimate or determination of how much of that loss will be offset by wetland migration.

Wetlands span most bay shorelines in Texas, and erosion is especially prevalent along north-facing bay shores where wave activity is vigorous during the passage of fronts when northerly winds persist for several days. A study of West Bay, located on the landward side of Galveston Island, showed that virtually the entire southern shoreline of the estuary is eroding at rates greater than 3 feet/year (White et al., 2002). In response to this rapid loss of wetlands, rock groins and sand-filled breakwaters have been placed along much of the shoreline. The northern shoreline of the estuary is rimmed by a narrow stretch of wetlands whose inland migration is blocked by spoil banks of the Intracoastal Waterway.

Wetlands of the central Texas coast are also being inundated and eroding at rates that in many locations are as high as 3 feet/year and locally
exceed 10 feet/year (White et al., 2002). These high rates are mainly the
result of wind erosion and low vertical accretion rates. Fringing wetlands
along the north shore of the Matagorda Peninsula provide a good example
(fig. 7.2). East Matagorda Bay averages 3.5 miles in width, resulting in a
modest fetch for northerly winds. Extensive fringing wetlands characterize
the north-facing shoreline. In contrast, the shoreline of the western half
of the peninsula, which is exposed to northerly winds that blow across
10 miles of open bay, is dominated by wave-eroded shallow sand flats.
Inland migration of wetlands along the north shore of Matagorda Bay and
East Matagorda Bay is restricted by dredge spoil banks of the Intracoastal
Waterway.

a

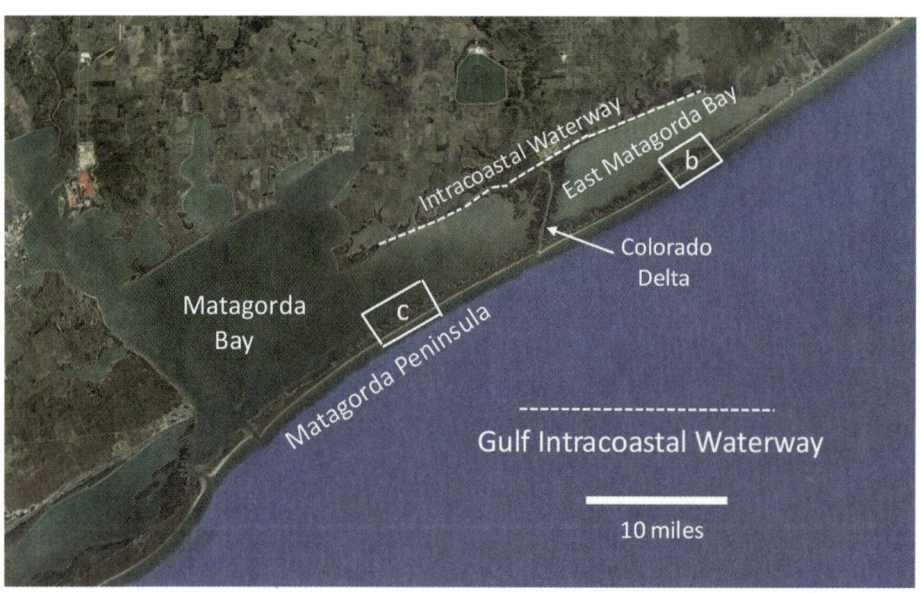

FIGURE 7.2. *Above and facing,* (a) Location map for the Matagorda Bay area.
The dashed line is the Intracoastal Waterway, lined by spoil ridges and the Ingleside
Paleoshoreline, which block wetland migration. (b) Image of East Matagorda Bay
showing fringing wetlands behind the Matagorda Peninsula. (c) Image showing
wave-eroded shoreline with patchy wetlands behind the western Matagorda
Peninsula. All from Google Earth.

b

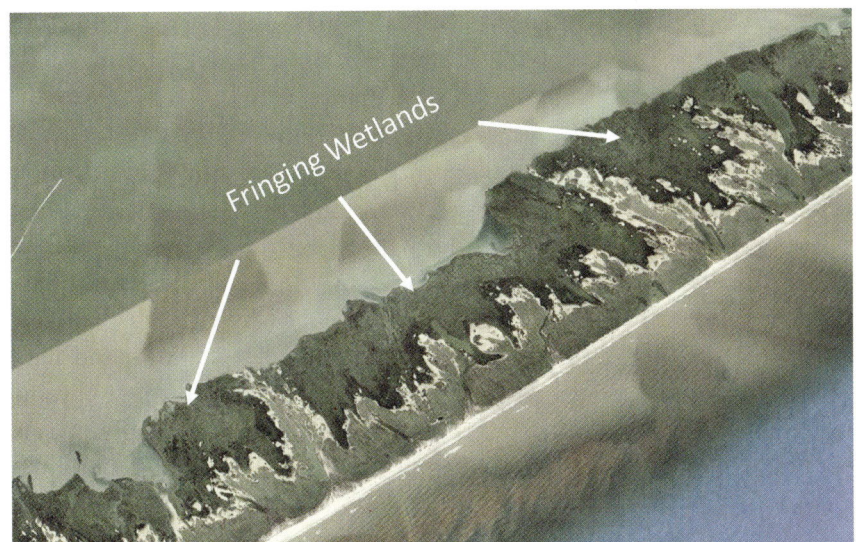

c

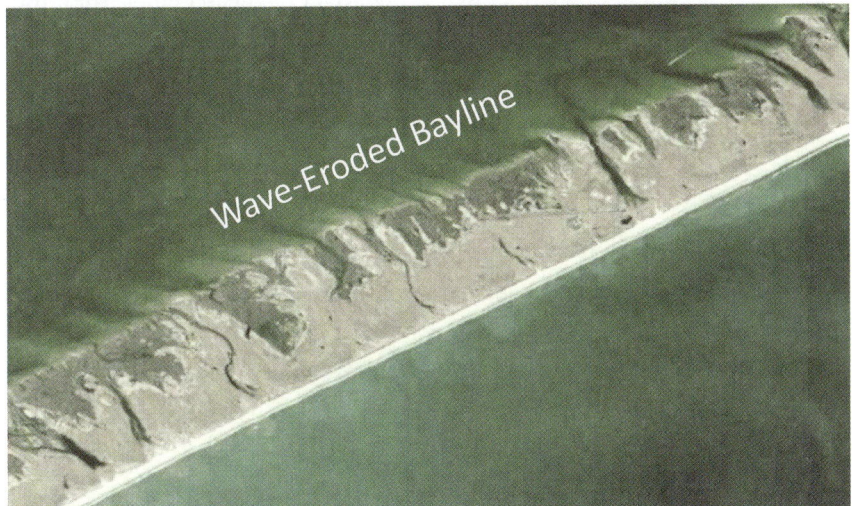

Mississippi, Alabama, and Florida Wetlands

The eastern coasts of Mississippi and Alabama, as well as the western
coast of Florida, are characterized by restricted fringing wetlands along
the shores of estuaries that occupy narrow, low areas bounded by relatively
steep topography and by developed areas. Thus, these wetlands have limited
opportunity for landward migration as sea level rises. Before the 1800s,

the Mississippi coast was lined by saltwater and freshwater wetlands. An estimated 60 percent of those wetlands have since been lost, primarily to coastal development. The most extensive remaining wetlands in Mississippi Sound occur along the northeastern side of the estuary between Pascagoula and Mobile Bay. Given a lack of historical data, assessing the ecosystem impacts of wetland loss on Mississippi Sound and the Gulf of Mexico is not possible, especially considering other anthropogenic and natural effects. What is known is that the seafood industry in the region has suffered from declining fish and shellfish populations.

The Mobile-Tensaw bayhead delta is the largest area of wetlands west of the Florida Big Bend region. Most of the delta plain is destined to be inundated by rising sea levels before the end of this century (fig. 6.18). Northward migration of the delta plain will be restricted by topography, so there will be a steady decrease in wetland area. The most extensive wetlands along the northwest Florida coast are associated with the Escambia, Perdido, Blackwater, and Choctawhatchee River bayhead deltas. They, too, are threatened by sea-level rise and bordered by higher relief. Elsewhere along this stretch of coast, narrow fringing estuaries have been overtaken mainly by coastal development.

Florida's Big Bend

Tide-dominated coasts include coastal areas where tides play a crucial role in coastal geomorphology, sedimentology, hydrology, and ecology by maintaining a sizable intertidal zone. This generally requires a large tidal range, but one important exception occurs along the northwest Florida coast, in the area known as the Big Bend (fig. 7.3). The Big Bend coastline stretches more than 150 miles from the mouth of the Ochlockonee River south to Hernando Beach, Florida.

Large portions of the Big Bend coastline are owned by the state of Florida and designated as wildlife management areas. The exceptionally gentle coastal gradient extends offshore to a gently sloping continental shelf that dampens offshore wave energy. There are no coastal barriers, and the coastal

FIGURE 7.3. *Facing,* (a) Image of Florida's Big Bend area. The brown area is exposed to tidal influence, including many tidal creeks. From Google Earth. (b) View of wetlands with a palm-covered island in the background. (c) View of a grassy Gulf shoreline with low wave energy, dominated by black needle rush.

a

b

c

plain is occupied by extensive marshlands with abundant tidal creeks and tidal flats that extend several miles inland from the Gulf shoreline (fig. 7.3a). These coastal environments rest on a surface with rugged karst topography cut into limestone. Holocene sediments are typically less than three to six feet thick, contributing to the region's low subsidence rates.

As a result of its broad intertidal zone, the Big Bend area is home to one of the more unique coastal ecosystems of the Gulf Coast. Sediment supply to the coast from rivers is minimal because local rivers, like the St. Marks River, are mostly spring fed, as reflected by clear water and grassy riverbeds. Inland areas are characterized by extensive marshes and lakes with small, palm-covered islands appearing like oases in the vast spans of wetlands (fig. 7.3b). The coast is mostly an intertidal zone with a narrow beach and grassy shorelines dominated by black needle rush (*Juncus roemerianus*) (fig. 7.3c).

Given its low, flat topography, Florida's Big Bend region is vulnerable to rising sea levels and storm surges. However, these impacts may be overshadowed by hydrological alterations of its watershed, leading to increased nutrient loading and decreased water clarity. This is already occurring in the form of a significant decrease in seagrass extent in recent decades. Unfortunately, the causes are hard to document and monitor because much of the region drains through a complex network of subsurface limestone cavities and channels. As a result, direct flow measurements are challenging, and there is a lack of historical measurements needed to assess how drainage to the coast responds to natural climate events and human alteration of the regional water cycle.

In addition to expansive seagrass loss in the Big Bend area, there has been a significant reduction in oyster reefs, which have existed in the shallow coastal waters for thousands of years. A study of changes in the areal extent of oyster reefs in the region between 1982 and 2011 found a 66 percent net loss of reef area. These losses occurred mainly offshore (88 percent loss), followed by nearshore (61 percent loss) and inshore waters (50 percent loss) (Seavey et al., 2011). This decline is attributed to reduced freshwater inputs from groundwater discharge to the coast, but again, this is hard to document. What is clear is that dryer conditions, as predicted in some climate models, could lead to decreased freshwater discharge and increased salinities that will influence ecosystems such as seagrasses and oysters, which in turn provide essential habitats for a range of other organisms.

The Florida Everglades

From the Big Bend area to the Everglades, the west Florida coast is characterized by relatively steep coastal plain topography, with restricted intertidal areas suitable for wetlands. By far, the largest area of wetlands on the west Florida coast is the Florida Everglades, with marshlands that stretch across ~3,800 square miles of the coast. Approximately 60 percent of the area is less than three feet above sea level and is vulnerable to sea-level rise this century.

The flat, low elevation of the Everglades reflects its unique geological setting, consisting of a thin layer (about three feet) of organic mud resting on a relatively flat surface that cuts into Pleistocene carbonate rocks. This surface was cut by waves ~1.0 ka and then exposed as the area experienced regional-scale droughts that lasted between two and four centuries. This geological setting and the region's unique climate result in poor surface drainage, frequent inundation during unseasonable tides, flooding during high rainfall, and exposure during droughts. This produces a diverse wetland ecosystem composed of highly tolerant saw grass (*Cladium jamaicense*) and heavily wooded areas separated from Florida Bay by mangrove swamps.

NOAA model results show widespread inundation of the Everglades by 2050 with a one-foot sea-level rise (fig. 7.4). The impact of sea-level rise is exacerbated by the relatively low vertical accretion rate of mangroves; drowning occurs when RSLR exceeds 0.2 inch/year (5 mm/yr) (Saintilan et al., 2020). The changes projected by the NOAA inundation model, coupled with severe droughts, peat loss, wildfires, and loss of the unique ridge and slough topography, are expected to cause significant shifts in plant and animal communities and increased exotic species invasions. Rising temperatures and high levels of nitrogen and phosphorus from agricultural runoff have already led to algal blooms that have depleted oxygen levels in Florida Bay, impacting freshwater and marine life. This is expected to worsen as atmospheric and Gulf water temperatures continue to increase, coupled with accelerating sea-level rise that is already taking its toll on mangrove wetlands that rim the coast, further increasing the threat from hurricanes (Hine et al., 2016). These historical changes sharply contrast with prehistoric times when mangrove-covered coastlines grew significantly seaward (Hine, 2013).

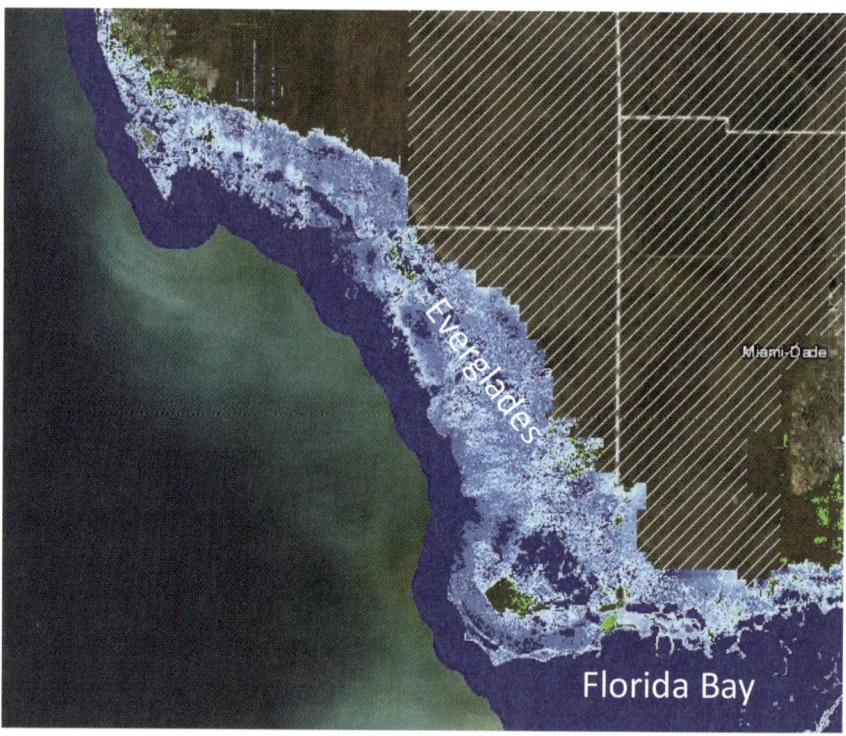

FIGURE 7.4. Model results from NOAA show the extent of submergence of the Everglades by the one-foot rise in sea level projected for 2050, shown in light blue. Dark blue designates areas currently below sea level.

The battle to save Gulf Coast wetlands has been lost. Most existing coastal wetlands will have vanished by midcentury, but the ecosystem impacts are already occurring. This outcome was initially the result of direct human activities, of which there is a long list, including treating wetlands as landfills where sprawling coastal communities, harbors, and industry are now located. But the final demise of wetlands will be caused mainly by rising sea levels. Areas most affected include virtually all of coastal Louisiana and the Florida Everglades. Still, the cumulative impacts transcend the Gulf Coast, as fringing wetlands have little space for migration as the sea level rises. Assessment of the ecosystem damage from wetland loss is impossible because of scarce historical data. But the cost of lost ecosystem services and hurricane impacts alone will likely be measured in trillions of dollars, and our children and grandchildren will pay the bill. This does not include damage from major oil spills.

I was introduced to the Big Bend area when I was a graduate student at Florida State University in Tallahassee. I was intrigued by the contrast between the region and the Alabama coast, where I spent my childhood and teenage years. Having since visited hundreds of coastal settings across the globe, I still consider it one of my favorite coastal areas. The drive to the coast from St. Marks, Florida, through the St. Marks National Wildlife Refuge gives one the sense of traveling back in time. The route first passes through a dense palm/palmetto forest that gives way to isolated oases of palm-covered islands rising above extensive marshlands. Farther south, one passes through networks of intertidal marshland with tidal creeks teeming with wildlife, including alligators. A walk along the beach or a trip to one of the area's observation decks is a must, and the St. Marks Lighthouse provides a beacon for a safe return to the parking area. On an average day, one can wade offshore or even kayak in the calm, clear Gulf waters to observe extensive seagrass meadows with incredible and unique sea life. I hope my grandkids will visit the area and that it will not have changed too drastically before then.

8

Severe Storms

Hurricane impacts pose the greatest uncertainty in predicting changes to the Gulf Coast. Some scientists have argued that the number of landfalls along the coast will increase this century. Although this theory remains controversial, there is a strong case for the size and magnitude of storms increasing because of warmer surface water temperatures. Another result will be wetter storms because of increased evaporation. These arguments are supported by a recent history of storms that intensified before landfall and increased storm-related flooding. Storm impacts will increase with continued industrial and urban development in coastal areas and the inundation of coastal barriers and wetlands resulting from rising sea levels. Background image from NASA

Tropical storms and hurricanes form when warm, moist air masses rise and are replaced by cooler air, which then warms and rises (fig. 8.1). As this process continues, storm clouds form and heavy rains occur. As the air rises, it rotates clockwise because of the Coriolis force caused by the Earth's rotation. Cyclonic winds increase in intensity as more warm air is fed into the system, starting as a tropical disturbance and then becoming a tropical depression at speeds between 23 and 39 miles per hour. As winds intensify, a tropical storm develops (wind speeds 40 to 73 miles per hour) and becomes a hurricane if winds exceed 74 miles per hour. As the storm intensifies, cyclonic circulation strengthens to form a storm eye that marks the location of the most intense conditions. Once a hurricane develops, its

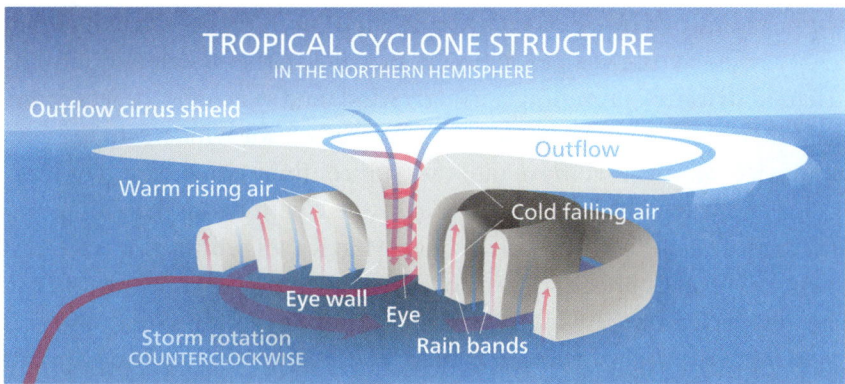

FIGURE 8.1. Hurricane development. Winds flow toward the storm's center because of the low pressure there, and air is forced upward. High in the atmosphere, winds flow away from the storm, which allows more air from below to rise, increasing storm intensity. From Kelvinsong/Wikimedia Commons.

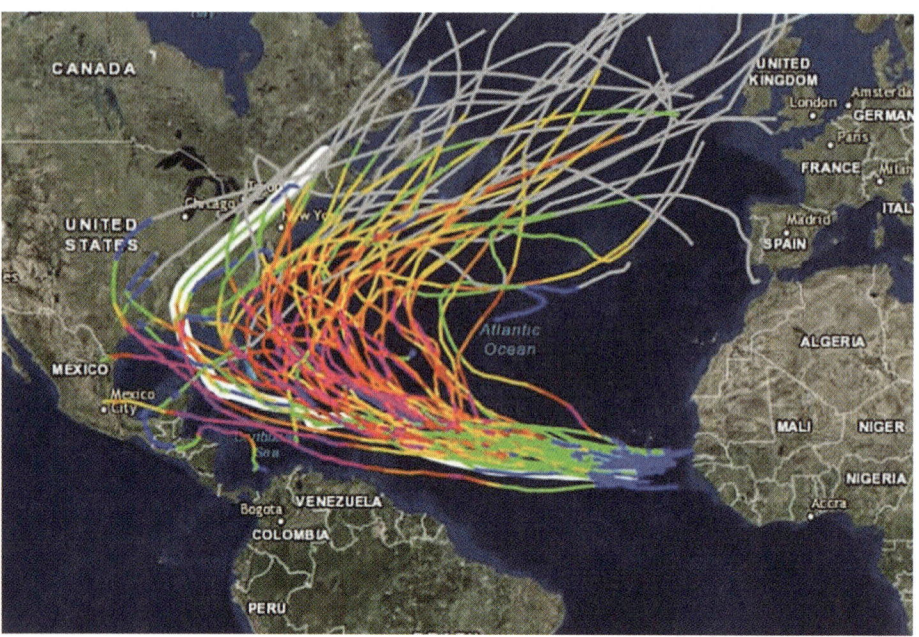

FIGURE 8.2. Hurricane tracks for the past 150 years. From the National Ocean Service, oceanservice.noaa.gov.

intensity is controlled primarily by sea surface temperatures and atmospheric water-vapor levels.

Hurricanes that threaten the Gulf Coast typically move west across the warm Atlantic Ocean, pushed by trade winds until they reach latitudes of about 25° to 30° (fig. 8.2). From there, they flow either north and into the Atlantic, driven by prevailing northerly winds and steered by the jet stream, or, if they continue to be influenced by westerly trade winds, into the Gulf of Mexico. Storms that move into the Gulf continue either west into Mexico or north toward the US Gulf Coast, their path being determined by low-pressure and high-pressure zones. Hurricanes will take a northeast trajectory if high pressure exists over the western Gulf. Low pressure in the western Gulf is conducive to storms impacting the Louisiana and Texas coasts. Hurricanes also form in the Gulf of Mexico, including two devastating storms, Helene and Milton, that made landfall along the Florida coast within two weeks of one another in 2024.

Once a hurricane enters the Gulf, its strength is determined by sea surface temperatures that fuel the storm and by high-elevation winds that may shear storm winds, robbing it of energy. There are also important oceanographic effects. For example, hurricanes can intensify significantly when moving across the warm surface water of the Loop Current (fig. 2.20). Increasing sea surface temperatures across the Gulf during the past few decades have provided fuel for more intense storms, including storms that have strengthened just before making landfall, a phenomenon known as rapid intensification. Warmer Gulf temperatures also equate to wetter storms because of increased evaporation and capacity for the atmosphere to hold more water.

Given the number of variables influencing storm magnitude and movement, forecasting storm paths and potential impacts is complex. Still, hurricane prediction has improved considerably with advances in our understanding of storm formation and direction, better computer power and forecast models, and improved aerial observations that allow meteorologists to determine whether a storm is intensifying or weakening. These advances have led to better prediction of hurricane tracks and landfall, decreasing evacuation times. Unfortunately, storm damage costs have steadily risen because of increased coastal development and impacts from flooding caused by high rainfall and *storm surges*, amplified by sea-level rise.

The Saffir-Simpson Hurricane Wind Scale classifies tropical storms and hurricanes. It was developed in the early 1970s by Herbert Saffir and Robert Simpson and was intended to provide a system for predicting the potential impacts of hurricanes. The original classification included five categories based on measured wind speed (table 1). Simpson later modified the initial

Table 1. Modified Saffir-Simpson Hurricane Scale

Category	Wind speed (miles per hour)	Atmospheric pressure (millibars)	Storm surge (feet above normal)
1	74–95	>979	4–5
2	96–110	965–979	6–8
3	111–130	945–964	9–12
4	131–155	920–944	13–18
5	>155	<920	>18

version to include storm surge. However, this approach has proven some-what inaccurate. The National Hurricane Center's *SLOSH model* has proven more reliable in forecasting storm surges, with an accuracy of 20 percent.

Storm Impacts

History has shown that storm impact increases over 400 times from a Category 1 to a Category 5 hurricane. The cost in dollars has increased as coastal populations and infrastructure have increased. Meanwhile, the loss of human lives has decreased thanks to better landfall predictions, building codes, and improved evacuation routes and procedures.

In his book *Coasts in Crisis: A Global Challenge*, Dr. Gary Griggs of the University of California, Santa Cruz lists five hurricane hot spots on the US East Coast and Gulf Coast that are especially vulnerable to hurricane impacts. Three of these are Gulf Coast cities, including Tampa–St. Petersburg, New Orleans, and Houston-Galveston. The vast majority of the most costly and deadly storms in the United States have made landfall on the Gulf Coast, with the 1900 Galveston Hurricane being the deadliest. That storm is credited with killing between 6,000 and 12,000 people and destroying over 3,000 homes in what was at the time an economic hub for Texas. The storm's winds exceeded 120 miles per hour, and the storm surge is estimated to have been between 8 and 15 feet. The second most deadly storm was the 1928 Okeechobee Hurricane, which claimed between 2,500 and 3,000 lives.

Since wind measurements have been recorded, there have been relatively few Category 5 hurricane strikes on the Gulf Coast. In 1992, Hurricane Andrew made landfall on the southeastern Florida coast as a Category 5 storm, destroying over 60,000 homes and causing record-breaking damage estimated at $50 billion. Since Hurricane Andrew, Florida's population has grown nearly 60 percent, and the number of new coastal residents has increased significantly, including people living on barrier islands and other areas with high storm impact potential. On October 10, 2018, Hurricane Michael made landfall near Mexico Beach, Florida, with winds of 160 miles per hour, resulting in an estimated $25 billion in damage and claiming 59 lives.

Six Category 4 hurricane strikes along the Gulf Coast have occurred since 2018. The most recent was Hurricane Helene, a massive storm that rapidly intensified to become one of the most powerful storms to strike

Florida's Big Bend region, with sustained winds of 140 miles per hour. The storm was fueled by record-warm waters and high amounts of water vapor, resulting in record rainfall and flooding that extended from Florida to the southern Appalachians. It followed on the heels of Hurricane Ian, which struck the southwest Florida coast just two years earlier, making landfall near Cayo Costa, Florida, with a maximum wind speed of 150 miles per hour and a storm surge of 15 feet. Ian resulted in at least 120 deaths and an estimated $110 billion in damage. In October 2024, Hurricane Milton formed in the western Gulf and rapidly intensified as it moved across the warm Gulf waters. All three storms were narrow misses for the highly vulnerable Tampa–St. Petersburg area.

Whether the recent trend is a sign of the increasing frequency of powerful storms impacting the coast remains uncertain. However, the number of storms undergoing rapid intensification, meaning wind speeds exceeding 35 miles per hour within 24 hours of landfall, has increased since 1982. This is consistent with climate models that indicate increasing intensification of tropical cyclones as a result of increasing water temperatures and relative humidity along with decreases in vertical wind shear (Balaguru et al., 2024). The most recent additions to the record books are Helene and Milton, which made landfall along central Florida within two weeks of one another in 2024. Hurricane Milton was the fifth hurricane to strike the Gulf Coast in 2024, compared to an average of two hurricanes to make landfall in the United States per year, with a record of six landfalls in one year.

Storm surge impact potential is high along the western Gulf Coast because of lower coastal plain elevations, the number of industrial facilities in Louisiana and Texas, and large population centers, including Houston and New Orleans. The most devastating storm surge came from Hurricane Katrina in 2005. Katrina was a massive storm with an estimated storm surge of 28 feet that left a broad swath of damage stretching from New Orleans to the Mississippi coast. New Orleans was flooded when the surge breached levees, spilling water across roughly 80 percent of the city from the swollen Mississippi River. Before Hurricane Katrina, the most significant recorded storm surge was associated with Hurricane Camille in 1969. Camille struck the Mississippi Gulf Coast with a record-breaking storm surge of 24 feet.

The record for the wettest hurricane was set by Hurricane Harvey in 2017. Harvey made landfall along the central Texas coast as a Category 4 hurricane, inflicting heavy damage. The storm then reversed course and circled back out over the unusually warm surface waters of the Gulf of Mexico, where it gained strength before making its second landfall. Harvey dumped record amounts of rain and caused extensive flooding in the Houston metropolitan area, resulting in a once-in-1,000-years flooding event in just one day and causing an estimated $125 billion in damage.

Climate Change Influence

A landmark paper by Dr. Kerry Emanuel (2005) provided statistical evidence that the duration and intensity of hurricanes worldwide had increased by about 50 percent since the 1970s. This trend directly corresponded with a global increase in tropical sea surface temperatures. More recently, there has been improved quantification of this effect, showing that for each 1°C (1.8°F) increase in surface water temperatures, wind speeds of major hurricanes increase by about 10 miles per hour, adding to their destructive potential (Mann et al., 2017). These observations indicate that the number of Category 4 and 5 hurricanes will increase globally, including in the Gulf of Mexico (Knutson et al., 2019). These will be wetter storms, so their impacts will extend farther inland. The atmosphere can hold about 7 percent more water for every 1°C increase in temperature. This could result in a 30 to 40 percent increase in precipitation for storms in the Gulf of Mexico (Bruyère et al., 2022). Thus, Hurricanes Harvey and Helene may be a harbinger of future storms.

Results from a recent weather-pattern statistical model show that increasing sea surface temperatures over the 1982–2020 period doubled the probability of "extremely active tropical cyclone seasons" (Pfleiderer et al., 2022). In 2020, there were 32 tropical cyclones in the Atlantic, the highest number on record, including five of the most destructive Atlantic hurricanes. Another prediction is that increasing aridity will result in greater evaporation and moisture content in the atmosphere.

The reality is that the influence of climate change on severe storms remains unclear, so we are "learning as we go," especially when it comes to how atmospheric changes might influence storm tracks. We know that storm tracks in the Gulf are controlled by large pressure systems that generally migrate from west to east across the Gulf Coast. It is becoming increasingly clear that the distribution and migration of these pressure systems are influenced by large-scale meteorological processes, such as the El Niño–Southern Oscillation (ENSO) and the North Atlantic Oscillation

(Rodysill et al., 2020). Historical records indicate that the variability in hurricane activity in the Atlantic Ocean has been influenced by multidecadal cycles in sea surface temperatures and atmospheric circulation changes associated with ENSO events. In general, El Niño conditions result in fewer hurricanes in the Atlantic and Gulf of Mexico because of stronger vertical wind shear and greater atmospheric stability, while La Niña favors warmer sea surface temperatures in the tropical Atlantic and lessens wind shear, enhancing hurricane activity.

One potentially positive impact of climate change is related to the influence of Saharan dust on storm development. Massive dust storms flow off the African coast from spring through early fall. These storms deliver superdry air to the middle of the atmosphere, producing a midlevel jet that creates vertical wind shear and impedes hurricane development (Dunion and Velden, 2004). In addition, the dust itself suppresses cloud formation by reducing the convection that delivers moisture to higher levels in the atmosphere (fig. 8.1). These effects increase as the size of the dust cloud increases relative to the size of the storm. Research suggests that increased temperatures in North Africa will result in more intense, and therefore more significant, dust storms.

Paleohurricane Records

Historical records of hurricane magnitude are too short to evaluate how the number and magnitude of hurricanes striking the Gulf Coast varied in the past. The record of hurricane tracks for the past 150 years shows wide variability, with the majority having remained in the eastern Gulf and Atlantic (fig. 8.2). Most of these storms occurred before detailed meteorological observations and airborne and satellite observations were available, so it is not possible to investigate how early storm tracks related to multidecadal (ENSO) climate cycles. Nor is it possible to observe long-term storm activity patterns and how they varied in response to changes in climate. As a result, scientists rely heavily on numerical models to predict how hurricane activity will vary in response to climate change.

The need for better long-term storm records has spurred a relatively new field known as *paleotempestology*, which relies on geological records of severe storm impacts. This approach involves collecting sediment cores from inland water bodies exposed to inundation and storm washover during storm surges. These studies have revealed thin sand and shell beds alternating with bay and lagoon mud (fig. 8.3). The sand and shell beds are interpreted as storm deposits composed of material eroded from beaches and deposited in these inland water bodies during storms. Radiocarbon dating provides a time frame for interpreting the frequency of storm events at a given location over thousands of years. Additional research has been aimed at analyzing the frequency and intensity of paleostorms using sand layer thickness, grain size, and distance traveled by storm deposits and calibrating this sedimentary archive to modern events (Wallace and Woodruff, 2020). This approach requires detailed knowledge of changes in coastal barrier locations relative to core sites and changes in the width and elevation of coastal barriers.

Results from several paleotempestology investigations are beginning to shed light on how hurricane activity varied in the past. Some of the first studies were conducted in the Gulf Coast region, including Lake Shelby, Alabama, and Western Lake, Florida. The results from these studies showed that hurricane impacts were relatively high between ~3.5 ka and ~0.7 ka and decreased during the last 700 years (Liu and Fearn, 2000a, b). These results are supported by a more recent study that focused on the record of intense hurricane landfalls in the Florida Panhandle during the past 2,000 years. These records show greater storm activity from 2.0 ka to 0.6 ka, followed by more quiescent conditions over the past six centuries (Rodysill et al., 2020). Thus, the combined results indicate relatively high storm activity during the late Holocene, roughly coinciding with the Medieval Warm Period, followed by a decrease during the Little Ice Age when atmospheric temperatures were the coolest they had been for 8,000 years (fig. 3.2).

A recent study of central Texas showed periods of increased tropical storm activity from ~4.4 ka to ~3.9 ka and from ~0.3 ka to the present, which is out of phase with records from Florida and Alabama (Monica et al., 2024). This is because the path of hurricanes is strongly influenced by the distribution of high- and low-pressure systems, which typically varies across the Gulf Coast. This further suggests climate control of hurricane movement and landfall, leading Wallace and his colleagues (2014) to argue that paleohurricane record compilations may be well suited for detecting Gulf-wide trends in tropical cyclone activity. This is supported by studies with larger geographic coverage, including Florida, the Caribbean, and the Bahamas (Lane et al., 2011; Wallace et al., 2014). These and other studies are yielding a clearer picture of these large-scale influences, including ENSO, the West African Monsoon, latitudinal shifts in the Intertropical

FIGURE 8.3. (a) A sediment core from Puerto Rico with storm beds composed of sand and shell debris. From Wallace et al. (2014).

Storm Deposit

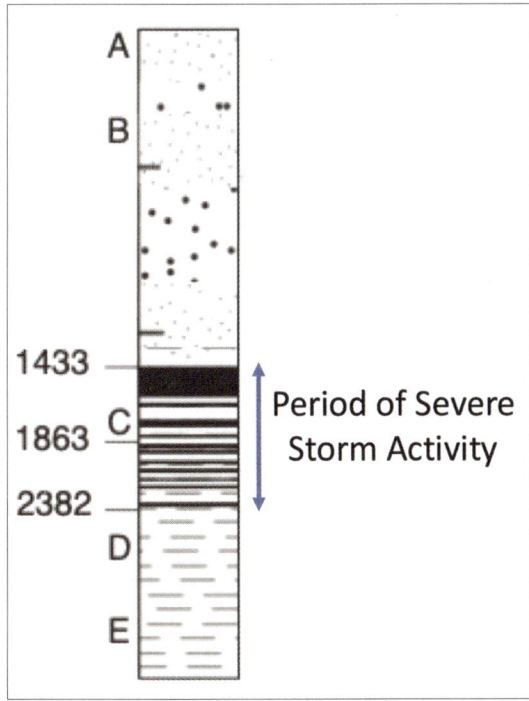

A

B

1433

C

1863

2382

D

E

Period of Severe Storm Activity

FIGURE 8.3. (b) An example of concentrated storm beds (black) within a sediment core from Laguna Madre. Shells include *Donax* and other mollusks that inhabit the Gulf shoreline. The core shows increased hurricane activity between 2382 and 1433 BP (blue arrow). The record ended about 900 years ago as washover sands from South Padre Island began to fill the lagoon. Multiple cores from different locations within the estuary were used to piece together a more complete record of hurricane impacts. Data from Wallace and Anderson (2010).

Convergence Zone, southerly shifts in the Loop Current, and shifts in the jet stream and Bermuda High to the south and southwest, resulting in warmer and dryer conditions in the Gulf Coast region. The Loop Current influence is especially strong, as it plays a key role in modulating sea surface temperatures and salinity. Furthermore, northward shifts in the Loop Current are known to result in increased cyclone precipitation in the northern Gulf of Mexico (Bregy et al., 2022). These results and ongoing research are improving our understanding of how hurricane activity may change in the coming decades.

Megahurricane Records

Some scientists have argued for the addition of Category 6 hurricanes to the Saffir-Simpson Hurricane Wind Scale. This new category would include storms with wind speeds reaching or exceeding 192 miles per hour. This is based on the occurrence of powerful cyclones in the Pacific region in recent years. It raises the question, is there evidence for more powerful storms in the past? Since the geological record of storm impacts spans thousands of years, what evidence is there for more powerful hurricanes in prehistoric times? Among the geological data available for addressing this question are geomorphic features formed during severe storms, including storm surge channels and washover fans that record episodes of barrier breaching (fig. 5.7). In general, these features are larger and more numerous than those formed in historical time, but does this mean that storms were more powerful or that the barriers were narrower and lower when they were breached?

In Texas, the relative timing of barrier breaching and washover events was determined in several locations by radiocarbon dating of beach ridges and storm washover deposits (Anderson et al., 2022). The results show that for the most part, storm breaching and washover occurred when coastal barriers were younger, lower, and narrower than they are today. However, in some cases these geomorphic features imply stronger hurricanes than those in historical times. The Bolivar Peninsula, Texas, provides an example of beaching and washover occurring during the early stages of barrier development when the peninsula was narrower and lower. But after the peninsula had grown to its current size, it suffered a significant erosion event that decapitated the barrier, completely removing its beach ridges. This storm event was followed by the growth of a younger set of beach

Progradation

FIGURE 8.4. Aerial image of St. Vincent Island, Florida, looking west and showing alternating small and large beach ridge sets formed during the progradation of the island. The more prominent ridges are higher and have less dense vegetation, hence their white color. Red arrows show where smaller ridges are truncated by more prominent ridges, recording episodes of shoreline erosion presumably caused by major hurricanes that occurred at a roughly 300-year frequency. From Google Earth.

ridges (fig. 5.13). Radiocarbon ages indicate that this event occurred between 800 and 600 years ago. Sediment cores from Galveston Bay document this event as a spike in dinoflagellates, indicating a dramatic increase in estuarine salinities that likely resulted from an increase in the estuary's connectivity to the Gulf of Mexico (Ferguson et al., 2018b). It took nearly three centuries for salinities to return to prestorm levels.

Additional geomorphological evidence for more severe storms can be found on St. Vincent Island, Florida, which has one of the most pristine and well-dated beach ridge sets on the Gulf Coast. The record extends back nearly 3,000 years and shows relatively continuous growth of the island punctuated by erosion of smaller beach ridge sets and the formation of rather large ridges (fig. 8.4). The more prominent ridges are interpreted as having formed following major storm events, which occurred about every 300 years. An earlier study of beach ridges between Gasparilla Island and Sanibel Island, approximately 500 kilometers south of St. Vincent Island, revealed similar alternations between smaller and larger beach ridge sets

(Stapor et al., 1991). Age control is less precise than for St. Vincent Island but also suggests that the more prominent ridges record megahurricanes that formed at roughly 300-year frequencies.

There is still a lot of uncertainty about how climate change will impact hurricane frequency and steerage. However, there is growing scientific consensus that hurricanes are becoming stronger and wetter because of increasing surface water temperatures in the Gulf. This is consistent with more powerful, wetter storms resulting from a historical increase in Gulf temperatures. Meanwhile, more and more people are choosing to live along the coast, and there has been little reluctance to build and expand large industrial facilities in coastal areas. The result is a rapidly escalating potential for devastating economic impacts in industrialized coastal settings. Meanwhile, accelerating sea-level rise and associated widespread loss of wetlands and erosion of coastal barriers are increasing the potential damage of severe storms. Based on the latest NOAA predictions, exposure to landfalling hurricanes along the western Gulf Coast, assuming the highest RSLR rates, will double by 2050 relative to the risk at the beginning of this century.

Historical records of hurricane activity are too short to understand how storm activity relates to climate. Still, geological records indicate century-scale variability in hurricane frequency, with the past six centuries having been a relatively quiet period in terms of hurricane impacts. There is also geological evidence for larger and more powerful hurricanes having caused greater damage to the coast than has occurred in historical times.

I grew up and lived most of my life on the coast and have experienced my share of hurricanes. My parents lost their weekend house in Dauphin Island, Alabama, to Hurricane Frederic in 1979. But that same storm also significantly damaged their home 60 miles inland. In 1995, my wife and I built a weekend house on Galveston Island, a mile inland from the coast. Our house was built to withstand 160-mile-per-hour winds and a storm surge of 14 feet. At the time, ours was the tallest house in our coastal community. The house suffered minor damage during Hurricane Ike, but as my wife commented on our first poststorm visit, it was "not as bad as after the grandkids had visited." During Hurricane Harvey, our home in Houston was flooded. It was the first time the area had flooded in recorded time. After Harvey, we took refuge at our house on Galveston Island while repairs were made to our Houston house. This story's moral is that hurricane impacts are not restricted to the immediate coast. The coastal impact zone extends tens of miles inland to include the most populated cities on the Gulf Coast. This zone increases as storms become wetter and sea-level rise accelerates. Sea level has risen about six inches since we built our house on Galveston Island, and nuisance flooding is more common. We had hoped to pass the house on to our daughter and grandkids, but it now seems unlikely they will enjoy time on the coast, as have we.

9

What Is Being Done?

Humans have occupied the Gulf Coast throughout the Holocene, but until recently, our presence was hardly impactful enough to record our existence. Still, the Gulf Coast changed considerably during that time. Initially, these changes were mainly in response to rapid sea-level rise. By 4,000 years ago, sea-level rise was negligible, and the shoreline was near its current location. Still, coastal change continued in response to climate change. In the past century, humans have assumed the primary role in coastal change. Our alteration of the global climate has resulted in an accelerated rate of sea-level rise that has not occurred for 7,500 years. Coastal impacts from rising sea levels are amplified by low

sediment supply to the coast and direct human alteration. Efforts to assess and combat this change are not keeping pace with the rate of change.

Throughout the first half of the 1900s, the population of the Gulf Coast grew sustainably. This began to change during the post–World War II era as the US economy strengthened and more people could live and retire on the coast. During that same period, there was considerable growth of industries that relied on coastal waters for various purposes, including the discharge of industrial wastewater and cooling water directly into coastal waters. Growth of the oil and gas industry exposed the Gulf Coast to a new range of impacts, including rapid subsidence and oil spills. Currently, 30 percent of the US population lives along the Gulf Coast. The population is projected to increase significantly by 2050, putting population growth on a collision course with climate change, rising sea levels, and human alteration of the coastal environment. Even more alarming, the Gulf Coast is rapidly losing its coastal ecosystems, with the economic effects from wetland loss alone accounting for hundreds of billions of dollars over the next several decades (Costanza et al., 2014).

During the 1990s, climate change and accelerating sea-level rise were still hotly debated and their effects on the coast largely ignored, even though significant change occurred across the Gulf Coast. By the early 2000s, the reality of human influence on global climate was becoming more widely accepted, although impacts on the Gulf Coast were still poorly documented. The exception is Louisiana, where the roles of accelerated sea-level rise and human alteration of sediment supply and dispersal to the coast are recognized as the leading causes of widespread land loss. Unfortunately, this acknowledgment comes too late to save most of the Louisiana coast. Other states have been slow in recognizing the causes and magnitude of coastal change and are just starting to address the problem, putting them on the same trajectory as Louisiana.

Combined Roles of Sea-Level Rise, Climate Change, and Human Influence

It is generally accepted that rising sea levels pose the greatest threat to the Gulf Coast. Sea level has risen as much since 1990 as in the previous 100 years, nearly 10 times the rate of the past 2,000 years. The rate is expected to increase as contributions from the Greenland and Antarctic Ice Sheets increase. Greenland is experiencing the highest rise in atmospheric temperatures in the past 1,000 years and is expected to contribute as much as 1.5 feet to global sea-level rise this century. The Thwaites Glacier in Antarctica shows signs of collapse that could result in sea level rising by at least a foot within decades (Rignot, 2021). Even without these ice sheet contributions, sea level will rise enough during the next three decades to cause significant drowning of the Gulf Coast (fig. 1.2). Predicted impacts of sea-level rise do not fully account for climate and human-induced changes in sediment supply to the coast or damage by more powerful hurricanes.

It is essential to recognize that the coastal change scenario shown in figure 1.2 is well underway and that these changes are a dramatic reversal from the long-term trend in coastal evolution. The most notable example is the Mississippi River Delta, which grew dramatically during the past few thousand years. Over decades, that growth has shifted to rapid land loss. Saving the delta is proving to be a monumental task that is expected to be met with minimal success. Meanwhile, excessive water usage in the wake of increasing aridity in the Rio Grande drainage system has been so severe that the Rio Grande delta has vanished in historical time. The river is no longer a viable source of the sediment needed to maintain coastal barriers and wetlands. Most of the Gulf Coast is lined by barrier islands and peninsulas that isolate estuaries and lagoons from the open Gulf and help reduce storm impacts. They are also undergoing a significant reversal from a long-term growth trend to landward retreat (Anderson et al., 2023). This includes several barriers facing unsustainable retreat and overwash rates, marking the final phase of their life cycles. Large portions of the Louisiana Chenier Plain and the Everglades have crossed a tipping point where vertical accretion of wetlands is exceeded by sea-level rise, resulting in the rapid loss of these vital components of the Gulf Coast ecosystem. The Everglades is also experiencing widespread human encroachment and alteration of its surface water drainage system that will require tens of billions of dollars to rectify. Development of the Mississippi Gulf Coast has resulted in the loss of about 10,000 acres of coastal wetlands, significantly reducing nutrient input to Mississippi Sound, the largest estuary on the Gulf Coast. The sound also has significantly altered salinity because of the construction of the Bonnet Carré Spillway, with measurable losses in shellfish populations. Florida's Big Bend area, with its unique coastal setting and ecosystem, is changing because of alterations in the region's freshwater drainage system.

Other effects are more localized but highly impactful. Among these is the widespread drowning of bayhead deltas, critical components of estuarine ecosystems. The largest of these bayhead deltas, including the Nueces delta in Corpus Christi Bay, Trinity delta in Galveston Bay, Mobile Bay delta, and Apalachicola Bay delta, will lose significant portions of their highly productive delta plains by 2050. Tampa Bay is at the top of the list of estuaries threatened by urban growth and expansion. Runoff from urban areas has significantly altered the estuary's water quality, amplified by a complex tidal circulation system and irregular bathymetry. Several other estuaries, including Mobile Bay, Galveston Bay, and Matagorda Bay, have been dumping grounds for toxic industrial waste that has accumulated for decades in bottom sediments. Resuspension of these contaminated sediments by dredging and storms poses a significant threat to human health.

Coastal Impact Assessments

The most widely recognized coastal impacts are those caused by catastrophic events, particularly hurricanes, major chemical and oil spills, and coastal flooding. These are generally evaluated based on economic, ecosystem, and public health impacts. Impacts from climate change and rising sea levels are more challenging to quantify, but efforts to assess these have recently increased. The USGS Coastal Change Hazards website (https://marine.usgs.gov/coastalchangehazardsportal/) is an interactive site with maps that show the relative vulnerability of the US coast to sea-level rise and hurricane impacts. Short-term and long-term susceptibility to sea-level rise is determined using the Coastal Vulnerability Index (CVI). The CVI is based on several variables: regional coastal gradient, tide range, wave height, relative sea-level rise, and shoreline erosion and accretion rates. Storm impact vulnerability assessments are based primarily on coastal topography, with the most vulnerable areas in Louisiana and Texas, where coastal elevations are the lowest. This approach is reasonable but fails to fully capture the domestic threat from storm impacts, such as population density in low-lying areas and evacuation strategies for moving people from storm-threatened areas to high ground. It does not scale potential impacts in port industrial areas between large population centers and the Gulf of Mexico. These are primarily in Texas and include Houston, Freeport,

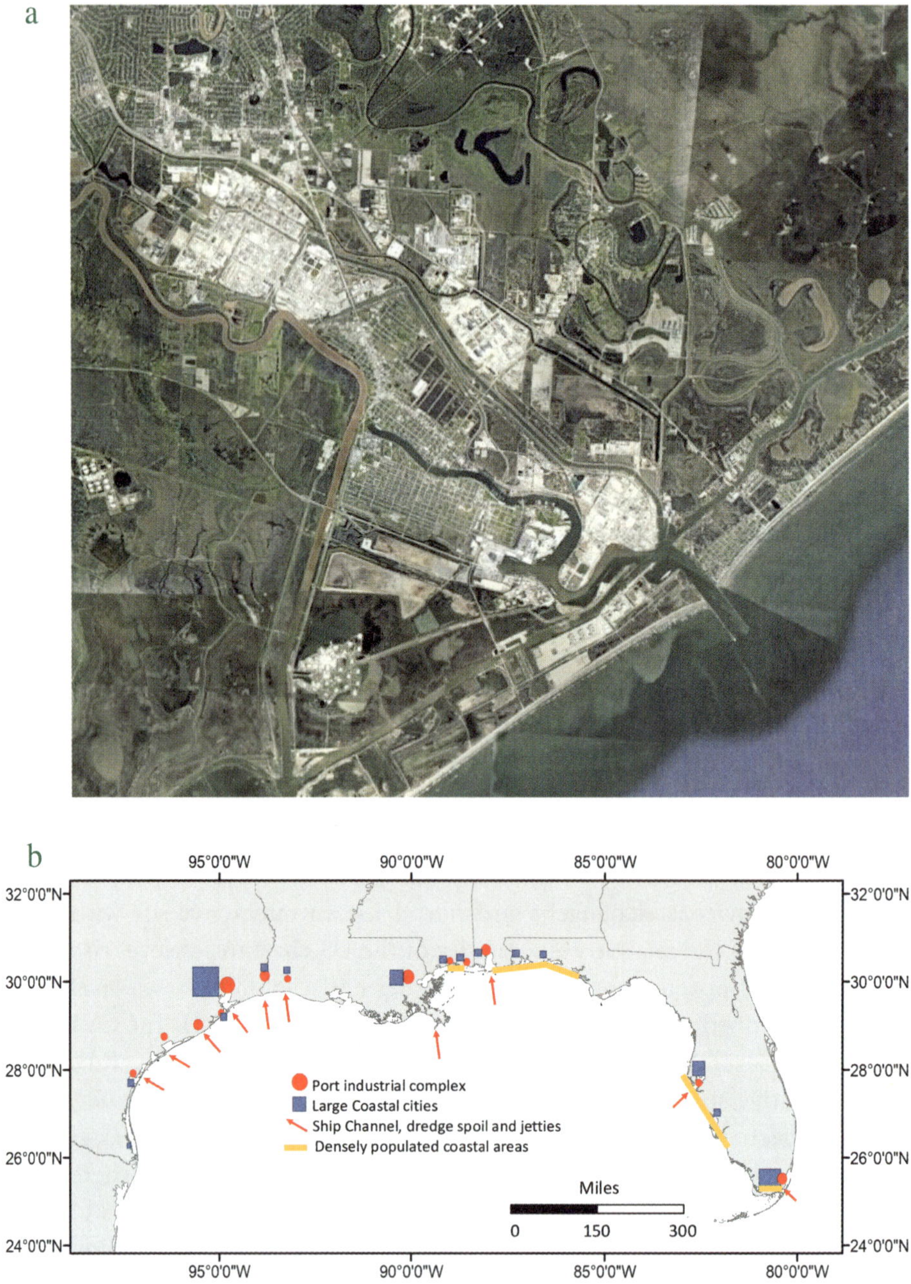

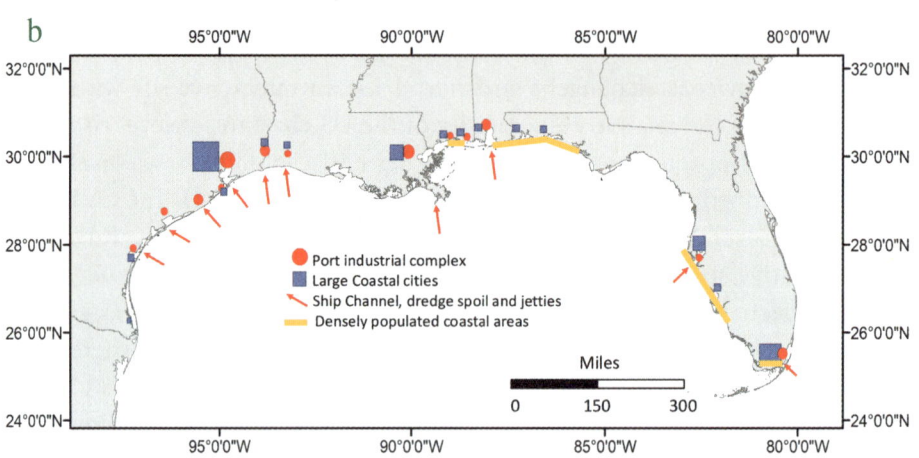

FIGURE. 9.1. (a) Aerial view of the industrial complex at Freeport, Texas, where a series of artificial levees provide the only protection from storm surge. From Google Earth. (b) Map of "high-risk" coastal areas.

Texas City, and Port Arthur, where large refineries and chemical processing facilities are located well below storm surge elevations and close to urban areas (fig. 9.1). A 2022 report issued by the Government Accountability Office found that 70 percent of facilities handling hazardous chemicals in Texas and Louisiana are vulnerable to sea-level rise, flooding, and storm surge. The entire Gulf Coast is under constant threat from oil spills. Other highly threatened coastal areas include large population centers where people live in low-lying areas threatened by storm surges. These include Houston, New Orleans, Miami, Tampa, and St. Petersburg. In just two years, the Tampa–St. Petersburg area narrowly escaped direct hits from Hurricanes Ian and Helene. Although the impacts of these storms were severe enough where they did make landfall, the effects of a direct hit on Tampa Bay would have been much worse.

Figure 9.2 shows areas with high potential impacts, based mainly on USGS coastal hazards maps. It shows much of the Mississippi, Louisiana, and Texas coasts with very high levels of vulnerability. In contrast, the Alabama and Florida coasts are classified mainly with high or moderate levels of vulnerability. One significant difference between these areas is the relatively high RSLR rates in Louisiana and Texas combined with low

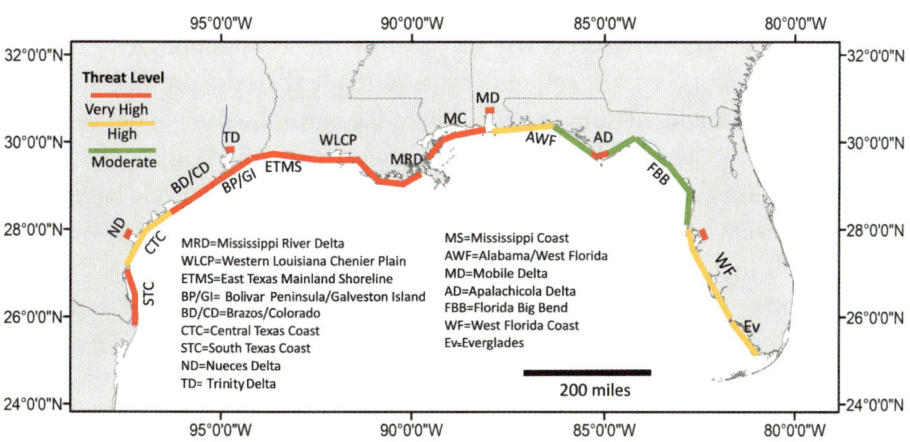

FIGURE 9.2. Coastal vulnerability map. It is based on rates of Gulf shoreline change measured by the US Geological Survey (data from Himmelstoss et al., 2017) and NOAA predictions for impacts of sea-level rise (fig. 1.2), as well as threats from storm impacts in port industrial areas and highly populated coastal areas with limited storm evacuation routes.

coastal plain gradients. The Miracle Strip of Florida is considered a high-risk area because of population density, limited storm evacuation routes, and increasing shoreline erosion rates. Coastal areas east of Panama City have lower vulnerability, except the Apalachicola River delta, whose delta plain is expected to be inundated by rising sea levels by 2050 (fig. 6.19). High population densities and limited evacuation routes threaten coastal areas between Tampa Bay and the Everglades.

Fortification of the Coast

Based on the information in previous chapters, it should be clear that the Gulf Coast is in severe decline. The causes of this decline are widely accepted by the scientific community. Unfortunately, politicians have not been as accepting of the problems, let alone the causes. The result has been delays in taking action or supporting solutions requiring more scientific scrutiny. The following are three examples of multibillion-dollar projects whose effectiveness remains problematic.

Saving the Mississippi Delta: The Classic Case of Too Little Too Late

When it comes to coastal change, Louisiana is in critical condition, being faced with the virtual demise of not only the Mississippi Delta but the majority of its coastal wetlands by 2050 (fig. 1.2). These problems are the result of decades of human alteration of the natural sediment delivery and dispersal system and accelerating sea-level rise. The Louisiana Coastal Protection and Restoration Authority (CPRA) currently leads large-scale restoration efforts on the Louisiana coast. Louisiana's Coastal Master Plan (CMP) includes several restoration projects to reduce land loss. Projects are prioritized and selected based on results from modeling and field observations. The state is investing $50 billion in the program.

One of the initial CMP projects focused on saving Plaquemines and the greater New Orleans area by creating breaches through the Mississippi River and Atchafalaya River channels to divert sediment into adjacent wetlands. One scenario called for two breaches near Breton Sound and Barataria Bay. It was estimated to result in about 330 square miles of land recovery for Barataria Bay. This plan was eventually scaled back to the Mid-Barataria Sediment Diversion Plan, which calls for a single breach that will divert sediment into Breton Sound (fig. 9.3). This plan is expected to save only about 20 percent of the wetlands remaining in Barataria Basin. The project received just over $3 billion in funding, with the construction phase beginning in 2023. Funding for the project comes mainly from the Natural Resource Damage Assessment from the Deepwater Horizon settlement and the National Fish and Wildlife Foundation's Gulf Environmental Benefit Fund.

FIGURE 9.3. Aerial view of the Mid-Barataria Sediment Diversion project. The project calls for breaching the levee and allowing sediment from the Mississippi River to be delivered to areas that are being inundated. It is estimated that it will recover only about 20 percent of the Barataria Basin's wetlands. From Google Earth.

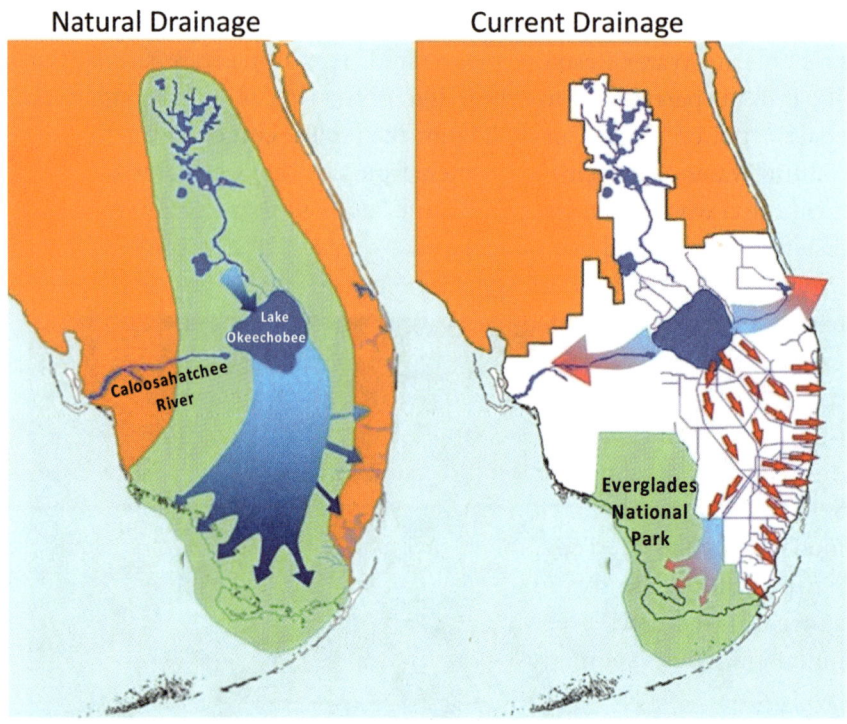

FIGURE 9.4. Changes to the south Florida drainage system and urban expansion have resulted in a roughly 50 percent reduction in the Everglades wetlands system. Credit: Everglades National Park Service.

Quest to Save the Everglades

Despite the designation of the Everglades as a national park and wildlife refuge in 1947, there has been an approximately 50 percent reduction in the spatial extent of the Everglades because of the alteration of south Florida's natural drainage system, coupled with urban encroachment. Under natural conditions, surface water drainage was directed mainly to the south into the Everglades (fig. 9.4). Drainage has been rerouted to Florida's southeast coast by canals, levees, and other flood control and water supply structures. It has also been rerouted to the west by dredging of the Caloosahatchee River. The overall intent of this massive alteration of the natural drainage system was to accommodate development and agriculture. The main impacts have been increased wetland salinities, decreased freshwater habitats, and widespread loss of seagrasses (Willard et al., 2011).

The Comprehensive Everglades Restoration Plan (CERP) was authorized by the Water Resources Development Act of 2000. It called for a federal-state partnership to restore and preserve the Everglades and its ecosystem. The US Army Corps of Engineers, in partnership with the South Florida Water Management District, is responsible for project planning. The project would eventually include 2,200 miles of canals and 2,100 miles of levees, berms, and other water control structures. The state of Florida and the South Florida Water Management District have invested about $2.3 billion in land acquisition, project design, and construction.

The CERP was finalized in 2000 and did not consider the effects of climate change, such as accelerated sea-level rise and changes in temperature, rainfall, and evapotranspiration. These issues were subsequently addressed by a workshop of experts convened by the National Academies of Sciences, Engineering, and Medicine. They used a regional hydrological model to predict 2060 conditions, including a temperature rise of 1.5°C, a ± 10 percent change in rainfall, and an 18-inch sea-level rise. The result was a scenario where sea-level rise had the most likely impact, with increased inundation in the southern portion of the park.

The Ike Dike Plan

Since the advent of steam-powered dredges and other heavy machinery, human response to coastal change has been to fortify the coast with seawalls, jetties, and other hard structures. The first significant project was the Galveston Seawall and associated jetties. Dredging of the channel between Galveston Island and the Bolivar Peninsula and construction of jetties on either side of the channel began in the late 1800s and proceeded in stages over several decades. Since their construction, the jetties have blocked longshore sand transport, resulting in as much as half a mile of beach expansion on both sides of the jetties. In contrast, the western half of Galveston Island was deprived of sand and experienced significant erosion (fig. 9.5a, b). After the "Great Storm" of 1900, the Galveston Seawall was constructed, along with a series of rock jetties that have disrupted longshore sand transport (fig. 9.5c, d). By the late 1950s, the island's west end had eroded, exposing the seawall to wave attack. The shoreline west of the seawall has since retreated about 400 feet from its 1950 location (fig. 9.5e, f). A shortage of sand needed for beach nourishment has prevented beach restoration west of the seawall.

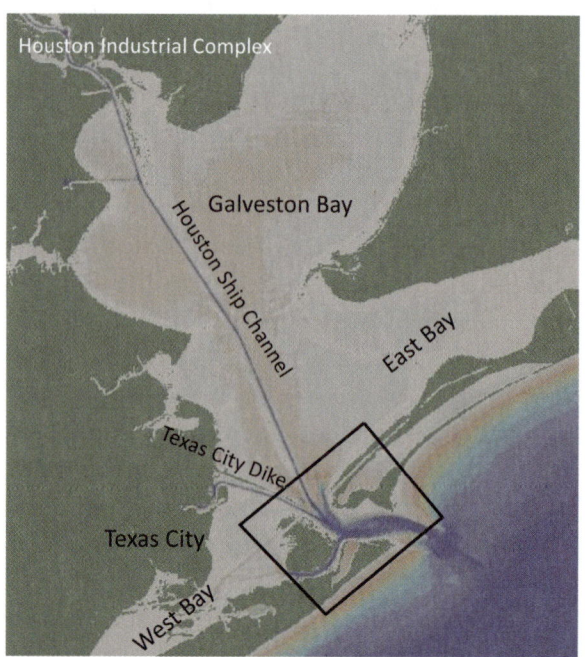

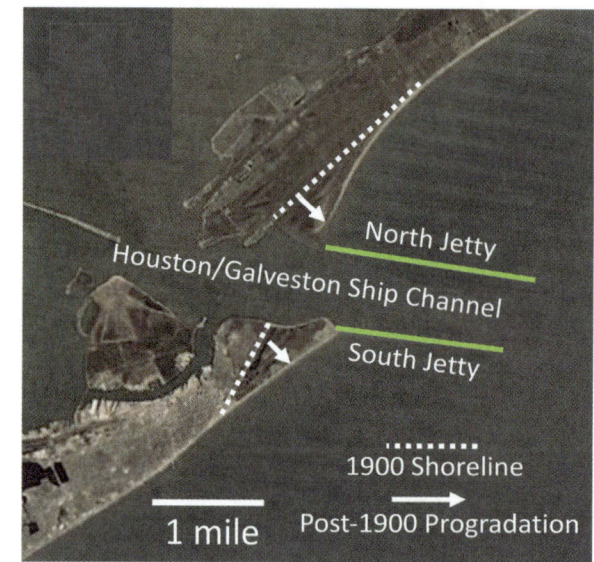

FIGURE 9.5. (a) The Houston/Texas City/Galveston Ship Channel. The box indicates the entrance shown in (b). From GeoMapApp. (b) Aerial view of the Houston/Galveston Ship Channel entrance with locations of the south and north jetties shown in green. The dotted white line shows 1900 shoreline locations. Note that significant shoreline accretion has occurred within one mile of the jetties. From Google Earth.

c

7312. Galveston Beach, Galveston, Tex.

d

FIGURE 9.5. (c) East Galveston Beach during the early 1900s. (d) 2005 aerial view of east Galveston Island shows sand accumulation on the east side of the rock jetties. This stretch of shoreline has since been nourished with sand.

e

f

FIGURE 9.5. (e) View from the western part of Galveston Seawall where the beach has eroded. (f) Aerial view of the west end of Galveston Seawall, where the beach west of the wall has shifted landward about 400 feet since 1950.

One of the largest proposed Gulf Coast engineering projects would occur on Galveston Island and the Bolivar Peninsula. This project became known as the "Ike Dike Project" because it was proposed following Hurricane Ike. This massive project intends to minimize storm surge in Galveston Bay and protect the coast from storm damage.

Planning for the Ike Dike was a joint effort between the US Army Corps of Engineers and the Texas General Land Office, with the Corps of Engineers doing most of the design. The original plan called for raising the Galveston Seawall from 14 to 21 feet and extending it with a series of levees and floodwalls for 43 miles along the coast (fig. 9.6a). The plan also called for constructing a series of large locks at the mouth of Galveston Bay to combat storm surge impacts in the bay (fig. 9.6b). A later addition to the plan called for constructing a 14-foot-high floodwall around Galveston. The estimated price tag for the project was $30 billion, and the time to complete it was roughly 30 years.

The Ike Dike Plan has changed repeatedly since its inception, and its name was later changed to the Galveston Bay Storm Surge Barrier System. These changes came partly in response to public outcry, which included concerns that the project involved too much physical infrastructure and little information on environmental impacts. In addition, there is concern that a study needs to focus more on the economic impact on Galveston Island and the Bolivar Peninsula during the many years of dike construction. The most recent plan has again been renamed and is now called the Enhanced Dune and Beach System. This new plan calls for constructing a 43-mile-long, 14-foot-high dune system and adding 250 feet of beach (fig. 9.6c).

The Corps "estimates" that construction of the Enhanced Dune and Beach System will require approximately 39 million cubic yards of sand. It is hard to say how this sand volume estimate was obtained, how much sand will be needed for beach nourishment, or where it will be obtained. What is clear is that it is not based on a detailed sand budget analysis. Such an analysis would be based on fieldwork that has not been conducted. One proposed borrow site is Sabine Bank (fig. 1.8), but a detailed study has shown that only a few feet of sand drape the bank (Rodriguez et al., 1999). Another proposed sand resource is the Trinity and Sabine River valleys, but studies have shown that the sand within these valleys is deeply buried in mud (Thomas and Anderson, 1989; Burstein et al., 2023). The environmental impacts of dredging the bank and valley are relatively unknown.

a

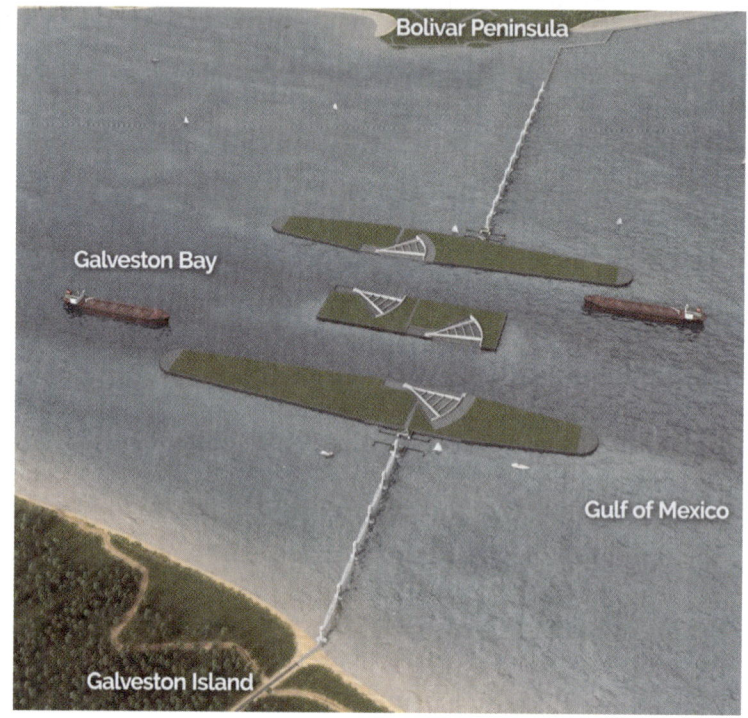

b

Bolivar Peninsula

Galveston Bay

Gulf of Mexico

Galveston Island

After

Before

c

FIGURE 9.6. *Facing*, The evolution of the Ike Dike concept. (a) An architectural rendering of the Ike Dike was posted in June 2015 by the Texas A&M University College of Architecture. From Archone e-news. (b) Sketch of floodgates at the Galveston Bay entrance as part of the project. US Army Corps of Engineers. (c) Revised project design (2022). US Army Corps of Engineers.

Construction of the Enhanced Dune and Beach System is not expected to be completed until 2050. By then, the coast will be drastically different from what it is today (fig. 9.7). Follets Island will have been reduced to a low, narrow barrier subject to more frequent inundation and washover, and the wetlands north and east of High Island will have been drowned, increasing the potential for storm surge from east and west of the surge barrier.

The Enhanced Dune and Beach System plan will likely continue to change, but that has yet to prevent it from reaching approval by the state of Texas and the US Congress. The project lacks the rigorous scientific oversight needed to evaluate its potential to prevent storm damage and environmental and socioeconomic impacts, to determine where needed sand resources exist, and to assess its long-term sustainability. It would be advisable for Texas to follow the lead of Louisiana and include more scientific oversight in coastal planning.

There are many other examples of human alteration of the Gulf Coast. An extensive inventory of these can be found in *Managing the Gulf Coast Using Geology and Engineering*, by Dr. Richard Davis and colleagues (2018). Unfortunately, there are no detailed accounts of the cost-effectiveness of these projects or their impacts on coastal environments. Gridlock remains between those who favor engineering solutions to coastal erosion problems and those who prefer more "natural" solutions, such as beach nourishment. Engineering solutions work only when they include environmental and economic impact assessment. These assessments must consider the impacts of sea-level rise and sand budget information essential for assessing how much sand will be needed for construction and beach maintenance.

The Search for Sand

Changes occurring along the Gulf Coast today are largely the result of a deficit in the sediment supply needed to keep pace with rapidly accelerating sea-level rise. This deficit has been caused by a long-term natural decline in sediment supply, exacerbated by human influences. Even the most optimistic projects for curtailing greenhouse gas emissions and rising sea levels offer

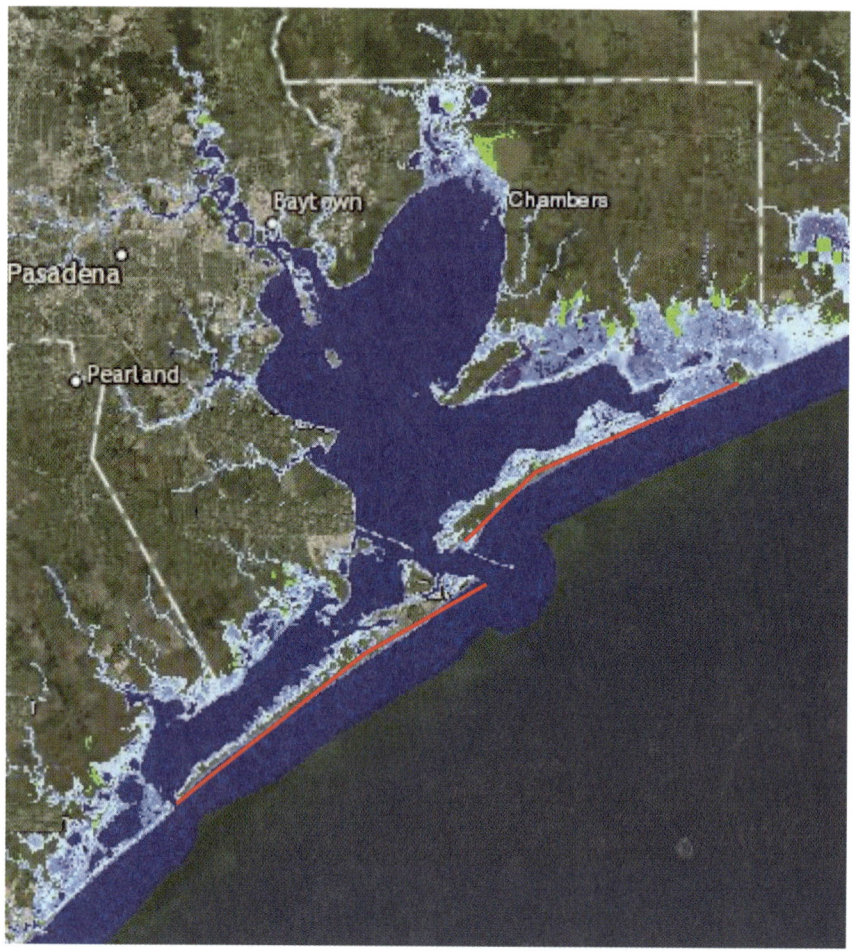

FIGURE 9.7. Model results from NOAA show the extent of coastal inundation by the projected one-foot sea-level rise by 2050, shown in light blue. Dark blue designates areas currently below sea level. The red line shows the location of the proposed Enhanced Dune and Beach System.

little hope for slowing widespread coastal inundation without substantial efforts to manage coastal sediment supply and dispersal. Recovery and maintenance of coastal wetlands requires nourishment projects with sand and fine-grained sediments. Better strategies are needed for dredge spoil disposal that do not restrict wetland and seagrass migration.

Unprecedented erosion is occurring on wave-dominated coasts, caused by a reduced sand supply and acceleration of sea-level rise exacerbated by decades of human alteration of sediment supply and dispersal systems (Anderson et al., 2023). Finding the sand needed for coastal nourishment is one of the most significant challenges in maintaining coastal barriers and mainland shorelines and requires a Gulf-wide effort. Louisiana is the only Gulf Coast state with a science-based, integrated initiative to measure rates of shoreline change, identify its causes, and search for the sand resources needed for large-scale shoreline nourishment projects. Louisiana's coastal barriers are under constant assault from hurricanes as well as increasing rates of sea-level rise, and they are struggling to survive. Significant and costly efforts have been made to reverse this trend by nourishing the islands. High rates of erosion on the mainland shorelines and chenier plain have been met with an aggressive beach nourishment program. These projects rely on sand acquired nearby that is accessible using existing dredging technology. These sand resources will diminish as the need increases.

Texas has a good record for monitoring shoreline change, and the results have shown that shoreline erosion is pervasive (fig. 5.12). However, more information is required to determine the fate of sand eroded from the coast, how much sand is required to stabilize the shoreline, and where the sand needed for large-scale coastal nourishment projects exists. Based on measured shoreline erosion rates, that volume is greater than what is known to exist nearshore. To the best of our knowledge, the only extensive sand resources within reach of conventional hopper dredges are the large deltas formed by the ancestral Brazos River, Colorado River, and Rio Grande deltas (fig. 9.8). Rough volume estimates indicate that these deltas contain about 40 million cubic yards of sand. Better volume estimates for sand suitable for beach nourishment will require a significant sediment coring operation. Seismic surveys have shown that the sand bodies associated with these deltas are close enough to the seafloor to be accessible by dredging, except on the central Texas continental shelf where the Texas Mud Blanket buries them. Other potential sand resources occur in old river channels on the continental shelf (fig. 1.6), but research has shown that these sand bodies are, with few exceptions, deeply buried in mud.

Using these offshore sand bodies to nourish the coast may be the most cost-effective solution for combating shoreline erosion in Texas. Given their distance from the coast, exploiting these offshore sand resources will be costly. I used to think it would be too expensive, but that was before the $30 billion Ike Dike Plan met with state and congressional approval. Given that this is the best, and really the only reasonable way to combat

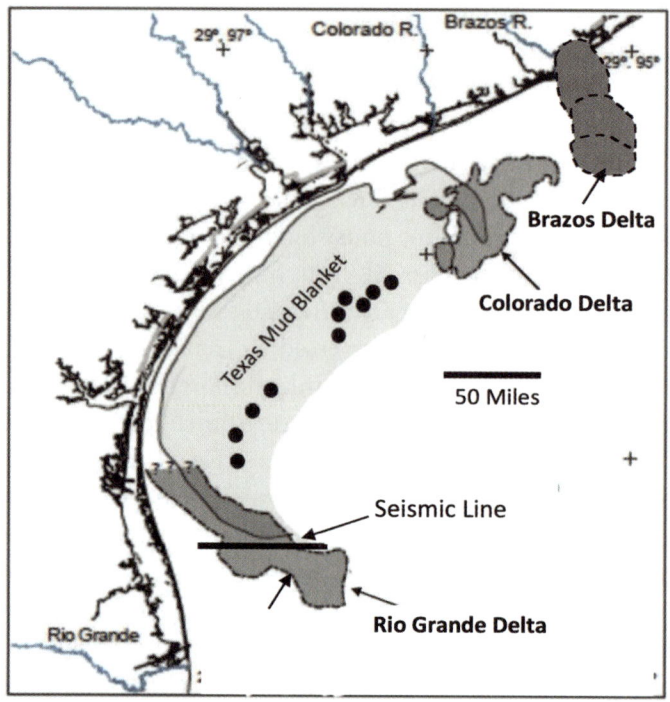

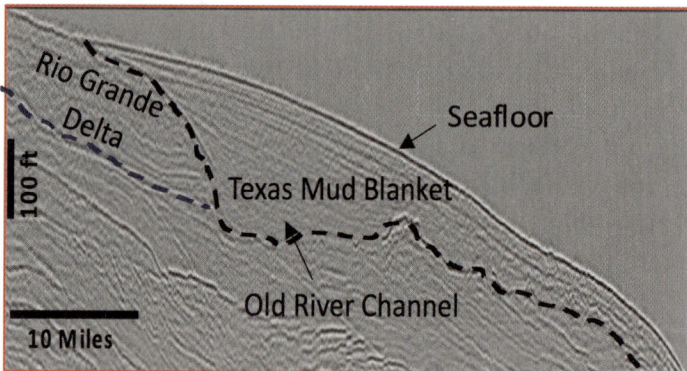

FIGURE 9.8. (Top) Map showing deltas on the Texas continental shelf
that are potential sources of beach nourishment sand. Black dots are
approximate locations of coral reefs, which once thrived on the central
Texas continental shelf but were drowned by sea-level rise. Modified
from Odezulu et al. (2020). (Bottom) Interpreted seismic section
showing the Rio Grande delta incised by a large river channel and
subsequently buried by the Texas Mud Blanket. Note that the delta on
the inner shelf is close to the seafloor. Modified from Weight et al. (2011).

shoreline erosion on the western Gulf Coast, there is a dire need to get on with quantifying sand needs and potential sources. The next step might be to commit a hopper dredge to long-range coastal nourishment full time.

The barrier islands of the Mississippi and west Alabama coast have all been migrating increasingly westward in historical time. Despite their distance from the coast, these islands provide protection from storm impacts. How changes to the locations of these islands and their areal extent could influence Mississippi Sound remains unclear, but sand nourishment will be necessary to decrease rates of change. Seismic surveys have revealed a complex array of river channels just below the floor of Mississippi Sound. Data collected seaward of the barrier islands show an extensive sand sheet likely formed by wave reworking of these fluvial sands (fig. 1.6). These sand bodies could nourish barrier islands and mainland beaches in the region.

Shoreline erosion is pervasive along the east Alabama and west Florida coastline. Fortunately, the MAFLA Sheet Sand covers most of the inner continental shelf between the Apalachicola River mouth and the Mississippi-Alabama border (fig. 1.7). It averages 10 to 20 feet in thickness, giving it a total volume of ~8.5×10^{10} cubic yards (McBride et al., 2004). That is enough sand to nourish Florida and Alabama beaches for decades, but additional mapping of the MAFLA using seismic data and sediment cores is needed. Offshore of Mississippi, the MAFLA transitions into sandy muds that are not particularly well suited for beach nourishment.

Any long-term strategy for beach nourishment needs reliable estimates of current and future erosion rates, and the latter hinge on a reliable sand budget that estimates sand supply and loss for different coastal segments. The federal agency responsible for identifying and assessing offshore sand resources is the Bureau of Ocean Energy Management (BOEM). It works with state agencies and universities in this effort, but progress in locating sand sources and establishing sand quality and volumes has been slow and must be accelerated.

States of Denial

The battle against coastal change has been largely reactive, with hundreds of billions of dollars spent on damage control after oil spills and major hurricanes. Several federal agencies, including NOAA, USGS, NASA, EPA, and FEMA, have focused on Gulf-wide issues, particularly sea-level rise and hurricane impact vulnerability. However, most Gulf Coast states need to be faster to recognize the full impacts of climate change on the coast and plan for future change. Indeed, the term "climate change" is commonly avoided by state agencies and even the media. Florida recently passed legislation prohibiting its use by state agencies. I once had a manuscript censored by the Texas Commission on Environmental Quality for using the term "climate change" and for providing evidence for accelerated sea-level rise. On June 1, 2024, Houston's ABC affiliate aired a special on the intensification of hurricanes in the wake of increasing Gulf water temperatures. No mention was made of climate change or the important role of rapid sea-level rise in storm surge impacts. Avoiding the term "climate change" only exacerbates our failure to recognize its reality. It is a bit like trying to educate the public on the dangers of smoking while avoiding the use of the word "cancer."

A significant step forward in combating coastal change was the passage of the Coastal Zone Management Act (CZMA), Section 309, amended in 1990 and 1996 to provide federal funding for voluntary coastal zone enhancement programs. Eligible programs include wetlands, public access, coastal hazards, cumulative and secondary impacts, energy and government facility sites, marine debris, ocean resources, special area management plans, and aquaculture. States are "encouraged" to conduct assessments of their programs every five years to remain eligible for funding. Oversight for the program is through NOAA's Office for Coastal Management (OCM).

Since Section 309 was amended, the progress by each Gulf Coast state has varied significantly, with Louisiana leading the way in science-based assessment of coastal change, its causes, and abatement plans. Unfortunately, Louisiana is also facing some of the most significant challenges and is having to make tough decisions in dealing with its vanishing coastal resources. In reality, most of the state's vast wetlands are unsustainable, and the impacts of their loss will reach far beyond the state's borders.

While the focus in Louisiana has been on wetland preservation and protection, additional efforts focus on barrier island restoration. An essential part of the barrier island effort has been sediment budget analyses for barrier islands and mainland shorelines. Since the 1990s, more than 40 barrier island projects have been completed using matching state and federal funds and proceeds from the Deepwater Horizon oil spill settlement. Among these are restoration projects on the barrier islands of the Barataria-Terrebonne Estuary and the Chandeleur Islands. The sand needed for these projects came from offshore. Locating these sediment resources was expedited by the state's efforts to synthesize existing data and acquire new data on offshore sand resources through the Louisiana Sand Resources Database (LASARD).

Texas has been lagging in understanding the causes and magnitude of its coastal problems, mainly because of failure to accept climate change and its impacts. The state's CEPRA (Coastal Erosion Planning and Response Act) aims to identify coastal areas needing protection and restoration. Early CEPRA publications in 2017 and 2019 were seriously deficient in scientific background and recognition of the impacts of climate change. The 2023 *Coastal Texas Protection and Restoration Feasibility Study* is a step forward in recognizing the historically unprecedented rates of landward migration of coastal barriers. It relies on the most recent NOAA sea-level predictions to estimate coastal change. The report also presents results from storm surge vulnerability models and includes detailed geohazard maps for the Texas coast. Unfortunately, the 2023 plan lacks a comprehensive sand budget analysis and sand resource database comparable to Louisiana's LASARD. There is also a paucity of subsidence data and marsh accretion data. Louisiana has hundreds of subsidence-measuring sites, while Texas lacks the sites needed to establish subsidence rates at the scale needed for restoration projects. Most of these are tide gauges in estuaries where Holocene sediments are thick, resulting in relatively high subsidence. Without reliable RSLR data, estimating the sustainability of proposed restoration projects is impossible. Fortunately, satellite data are starting to fill the gap in our knowledge of coastal subsidence, but there remains a paucity of the gauging data needed to calibrate satellite data.

Most of Texas's proposed restoration projects are geographically restricted, with a total estimated cost of $1.87 billion. The main exception is the Enhanced Dune and Beach System, including the surge gates at the mouth of Galveston Bay. It is currently estimated to cost just over $30 billion and is intended to protect about 16 percent of the Texas coast. The project will take 20 to 30 years to complete, and the price tag is expected to increase. The plan has evolved into a large-scale beach nourishment project requiring tens of millions of square yards of sand, leaving little sand for nourishing other coastal areas. A statewide assessment of future sand requirements is needed, as well as information on where sand resources exist. Results from offshore seismic surveys and sediment cores indicate that significant (millions of cubic yards) sand resources only exist far offshore, so exploiting them will be costly (Anderson et al., 2004, 2022).

My main criticism of the 2023 *Coastal Texas Protection and Restoration Feasibility Study* is that it focuses on changes that will occur by 2100, given sea-level scenarios of 1.6 feet and 4.9 feet, which are NOAA's intermediate-low and intermediate-high scenarios. The message here is that Texas has plenty of time to deal with its coastal issues. This is not the case. I use NOAA's one-foot-rise-by-2050 scenario in this book because it shows that significant coastal change will occur by that time. Thus, this 2050 scenario conveys a more realistic sense of urgency. Combating the magnitude of change by 2050 requires action now, and Texas is ill prepared to meet the challenge.

The barrier islands of Mississippi are eight miles distant from the mainland on average but still protect the mainland from hurricanes. The islands are not accessible by highways and are protected as part of the Gulf Islands National Seashore. In addition, the most populated coastal areas have a relatively steep coastal plain gradient, limiting the inland reach of storm surges. There are several escape routes for people living in the most populated regions of the coast. The state has had a long history of coastal research that continues today, led mainly by the University of Southern Mississippi. This research yields important information for locating offshore sand resources and conducting sand budget analyses.

The Mississippi Department of Marine Resources produces five-year coastal plan updates. The most recent assessment was for 2016–2020, with a draft assessment recently generated for 2021–2025. Progress has been slow in addressing the impacts of climate change and sea-level rise, despite NOAA's assessment of very high coastal vulnerability to sea-level rise in areas occupied by the state's remaining wetlands. There is also a paucity of the subsidence data needed to predict coastal change and help plan abatement projects. On the positive side, there appears to be abundant offshore sand for beach nourishment.

Alabama has only recently begun to monitor coastal change, but it has a long history of coastal research on which to build. Much of it has been focused on biological sciences, and relatively little work has focused on future climate change and rising sea levels. The Mobile-Tensaw bayhead delta will experience significant inundation in the next few decades, and the potential impact on the estuary's ecosystem is largely unknown (fig. 6.18). In addition, Dauphin Island has suffered a significant reduction in width and elevation, and the island's west end is struggling for survival.

The eastern coast is densely populated from Gulf Shores to the Florida state line, increasing the potential impact from storms.

The Alabama response to the Section 309 Enhancement Grants Program report avoids discussing the impacts of climate change and rising sea levels. More scientific information is needed to address coastal issues, their causes, and potential solutions. There are no comprehensive estimates of shoreline erosion rates, estimates of the volumes of sand required for beach nourishment, or information on where sand resources might exist. The demise of the Mobile-Tensaw bayhead delta and its economic and ecosystem impacts are not addressed. Based on this level of progress, the state is unprepared to maintain its rich coastal resources.

With its gifts of a warm climate, beautiful beaches, and emerald waters, no other state has benefited financially from its coast as much as Florida. Florida is home to unique coastal settings and ecosystems, including the Big Bend area, the Everglades, and the Florida Keys. Unfortunately, Florida's coastal environments are changing rapidly. State agencies have ranked most of the Gulf Coast shoreline as "critical erosion areas," including several high-risk coastal areas. These include coastal areas with a long history of stability and growth lasting thousands of years. The greatest threats to the Florida coast are rising sea levels and hurricanes, exacerbated by development practices that have included filling wetlands for residential and commercial development. As recently illustrated with Hurricanes Ian and Helene, the rapid intensification of storms puts more lives at stake in what is already a bad situation. Now, according to a recent analysis by Moody's, Florida faces the greatest economic risk of any US state.

Florida was one of the first Gulf Coast states to recognize and address coastal problems. The Florida Coastal Management Program involves several state and federal agencies that have successfully implemented several statutes to protect and enhance the state's coastal resources. Major projects are underway to restore the natural drainage system of the Everglades as well as its wetlands, and to protect Apalachicola Bay from further degradation. Both projects have relied on scientific input from state and federal agencies and academic experts. The state has also implemented several beach nourishment projects, depending mainly on offshore sand resources. Unfortunately, this history of science-based coastal management now faces political resistance. The state recently passed a bill prohibiting the use of the term "climate change" by state agencies. It remains unclear how this will impact efforts to preserve Florida's coastal resources, but it is clearly a deterrent to scientific communication. Unfortunately, Florida is not the only state to discourage scientific communication on the impacts of climate change.

Geological studies of the response of coastal environments to climate change and sea-level rise provide a record of dramatic and rapid coastal change. These changes occurred when rates of sea-level rise were similar to those occurring today. The main differences are that sediment supply was greater in the past, and humans were not a factor. After decades of climate change denial, we find ourselves in a situation where coastal change is occurring faster than expected, and the price will be the degradation and even demise of some of our most valued coastal settings, including much of the Louisiana coast and the Florida Everglades. The Gulf Coast's larger bayhead deltas will undergo widespread inundation by 2050, which will result in ecosystem impacts that remain poorly understood. Meanwhile, barrier islands and chenier plains are experiencing a reversal of the long-term trend of stability and growth to one of erosion and landward retreat. The debate about hurricane impacts is mostly over; storms are getting wetter and intensifying rapidly before striking the coast. This is especially problematic because large portions of the coast are highly vulnerable to storms, being densely populated and home to large industrial centers.

For the most part, the changes that will occur all along the Gulf Coast in the next few decades are irreversible. Even if we find ways to remove greenhouse gases from the atmosphere, that will not result in rapid cooling of the oceans or a reversal in the glacial and ice sheet contributions to sea-level rise. The only course of action for slowing the rate of rising sea levels is to reduce greenhouse gas emissions. Thus far, progress in that direction has been slow. Given this scenario, there is a dire need for improved coastal management to restrict population growth in areas threatened by hurricanes and rising sea levels. This will require a better understanding of how different coastal environments respond to rising sea levels and climate change. This research should follow the Louisiana model, where good collaboration exists between government agencies responsible for coastal protection and academic institutions engaged in basic research. These efforts should be coordinated and monitored by oversight panels of scientists and engineers with expertise in various coastal issues. More experts in coastal science will be needed, and academic institutions will need to expand their educational programs in coastal research and management. More public education on the realities of climate change and its impacts is also needed. This will require more funding, and we should rely on something other than another significant oil spill settlement to make that happen. The Gulf Coast should be treated as a vital part of the nation's infrastructure because sustaining coastal resources is critical to national security, economic prosperity, and the protection of living resources. Ultimately, the actual magnitude and rate of coastal change will depend on how well we can reduce greenhouse gas emissions, and thus far, we are not doing a good job.

After decades of attending countless meetings and reading reports on coastal issues, I am convinced that if the impacts of climate change and sea-level rise could be mitigated with acronyms, flow charts, data tables, and results from stakeholder surveys, we would be well on our way to solving our coastal problems. It is time for politicians to listen to scientists and accept the realities of climate change and its impacts.

GLOSSARY

aggradation: Increase in land elevation from the deposition of sediment.

aggradational barrier: A coastal barrier that grows vertically to maintain an elevation above sea level.

albedo effect: The positive feedback whereby a decrease in ice extent, including sea ice, results in greater heat uptake by the ocean.

alluvial valley: The valley occupied by a river over long time spans. It may include multiple channels packaged with floodplain deposits.

annual landfall probability: The time-averaged representation of hurricane impacts over a year, expressed as a percentage (i.e., number of hurricane events divided by years and multiplied by 100).

antecedent topography: The relict topography on which coastal environments rest.

Anthropocene epoch: A proposed geological epoch dating from the commencement of significant human impact on Earth's geology and ecosystems.

Arctic amplification: Changes in atmospheric temperatures and circulation patterns in lower latitudes that result from increasing atmospheric temperatures in the Arctic.

astronomical tides: Tides that result from Earth's rotation and the gravitational effects of the Earth, sun, and moon.

axial precession: Gravity-induced change in the rotational axis of the Earth.

backshore: Landward portion of a beach that extends from mean high-tide level to the base of dunes, cliff faces, or vegetation if dunes are not present. This portion of the beach is covered by water only during storms and exceptionally high spring tides.

back-stepping: A style of stratigraphic architecture characterized by landward shifts in sedimentary facies.

barrier island: An island composed of sand and separated from the coast by a lagoon or bay.

bayhead delta: A depositional landform formed by river discharge and sediment progradation into the head of an estuary.

bayline: The shoreline of an estuary.

beach ridge: A sand ridge constructed by erosion of the lower beach and deposition on the upper beach during storms.

breaker zone: The offshore location where waves break.

channel avulsion: A shift in river channel location.

chenier plain: A low-gradient coastal area composed of sandy or shelly ridges (cheniers) alternating with marshes and mudflats.

chenier ridge: Sandy and/or shelly ridges (cheniers) separated by marsh deposits.

clinoforms: Seaward-dipping strata formed by delta progradation.

coastal marsh: A low-lying wetland.

coastal plain: A low-gradient platform composed of seaward-dipping sedimentary deposits. The Quaternary coastal plain occurs within a few tens of kilometers of the coast and has gentle topography that is largely a product of wave erosion.

coastal sand budget: Sand supply to a segment of the coast by rivers and offshore sources, versus sand transported offshore by storms or landward by storm washover.

coastal wetlands: Low-elevation areas permanently or seasonally inundated with fresh, brackish, or saline water. They are inhabited by vegetation adapted to variable salinities and soil conditions.

continental shelf: The offshore extension of a continent characterized by its gentle gradient and water depths of less than 400 feet.

corrugations: Linear bedforms that occur in washboard-like sets at the grounding line of a marine ice sheet. Their width and height are consistent with their formation being influenced by rising and falling tides.

deforming bed: The mobile sedimentary bed beneath fast-flowing glaciers and ice streams.

delta headland: A protruding shoreline at the mouth of a river.

delta lobe: A lobate accumulation of sediment that forms where a river empties into a large body of water.

delta plain: The subaerial portion of a delta that includes distributary channels, natural levees, swamps, and interdistributary bays.

dinoflagellates: Single-celled organisms used as a paleosalinity indicator.

distributary channel: A channel that branches off a main river channel and is typically associated with river deltas.

Donax: A mollusk (bivalve) that inhabits the wave swash zone.

Doomsday Glacier: The unstable Thwaites Glacier in West Antarctica.

drowned river-mouth estuary: An estuary formed by flooding of a river valley.

ebb-tidal delta: An accumulation of sediment offshore of a tidal inlet transported and deposited during ebb tides and modified by waves and longshore currents.

ebb tide: Falling tide.

eccentricity: Changes in the shape of the Earth's orbit.

El Niño: The warm phase of the El Niño–Southern Oscillation associated with a band of warm ocean water that develops in the central and east-central equatorial Pacific.

estuary: An inland water body protected from the full force of offshore waves and storms by barrier islands and peninsulas, with variable salinity regulated by tidal flow through one or more tidal inlets.

eustasy: Worldwide change in sea level resulting from a change in the volume of ocean basins, changes in ocean temperature, or changes in the size and volume of ice sheets and glaciers.

evapotranspiration: The combined processes that move water from the Earth's surface into the atmosphere, including evaporation and transpiration.

flood-tidal delta: An accumulation of sediment, mostly sand, that is transported into an estuary or a lagoon during flood tides.

fluvial-dominated delta: A river delta where sediment delivery and dispersal are dominated by a river or distributary channels of a river, with little influence from tides and waves. Sediment supply from the river is large enough to allow seaward growth that results in bifurcating distributary channels that nourish individual delta lobes.

fluvial incision: The process by which rivers cut down to form deep (incised) river valleys. In coastal areas this downcutting occurs mainly in response to falling sea level.

foraminifera: Single-celled protozoans that form a test (shell) from calcium carbonate or by cementing sediment particles. They include planktonic forms that are useful paleotemperature indicators, and benthic forms that live at or near the sediment/water interface. Benthic foraminifera are relatively common in estuaries, where they can be good paleosalinity indicators.

foreshore: The intertidal portion of a beach that includes the swash zone.

fringing wetland: A wetland along the margins of an estuary.

geodetic benchmark: An elevation marker used to calibrate tide-gauge records.

glacial isostasy: Lithospheric depression or rebound resulting from the weight or melting of glacial ice.

glacial isostatic adjustment: Land elevation change caused by addition and removal of ice sheets.

glacial isostatic anomaly (GIA): Rebound of the Earth from melting and retreat of ice sheets that results from upward movement of mantle material.

glacier mass balance: The difference between glacier and ice sheet growth versus ice loss.

Global Positioning System (GPS): A satellite-based navigation system that provides accurate measurement of both geographic location and elevation.

greenhouse conditions: Climate conditions that occur during interglacials.

grounding zone wedge: A linear wedge of sediments formed at the grounding zone of an ice sheet.

ground-penetrating radar (GPR): A radar instrument used to image subsurface strata.

Gulf Loop Current: A surface current that transports warm waters from the Caribbean Sea through the Yucatán Channel and into the eastern Gulf of Mexico.

headland: A seaward protrusion in the coastline.

highstand: A period of maximum sea level when continental ice sheets were either absent or limited in size.

Holocene epoch: Geological epoch that spans the last 11,650 years.

iceberg furrow: A groove in the seafloor formed by an iceberg plowing across the bottom as it drifts.

icehouse conditions: A period in Earth's history when ice caps and ice sheets occur.

ice stream: Concentrated drainage from an ice sheet as it flows toward the coast.

Ingleside Paleoshoreline: The Gulf Coast shoreline of the last interglacial, approximately 120,000 years ago.

Intertropical Convergence Zone (ITCZ): A narrow zone near the equator where northern and southern air masses converge.

IPCC: Intergovernmental Panel on Climate Change.

jetty: A human-made feature, usually composed of rocks and constructed perpendicular to the shoreline.

lagoon: A landlocked coastal water body.

Last Glacial Maximum (LGM): The time when glacial ice on the Earth was last at its maximum volume, approximately 21,000 to 18,000 years ago.

LiDAR: Laser ranging instrument using GPS technology to measure inch-scale elevation changes.

Little Ice Age: A period of cooler conditions in the North Atlantic, and perhaps globally, that began in 1300 and lasted until the mid-1800s.

longshore current: Shore-parallel current in the nearshore zone that is produced by waves approaching the beach obliquely.

longshore transport: Wave-driven, shore-parallel sediment movement generally confined to the surf and breaker zone.

lower shoreface zone: An offshore area that is deeper than the fair-weather wave base but exposed to storm wave influence.

lowstand: Sea level at its lowest elevation.

MAFLA Sheet Sand: A sand body on the continental shelf off Mississippi, Alabama, and west Florida.

mainland beach: A beach that directly connects to the mainland.

mean rate of sea-level rise (MSLR): The average rate of sea-level rise over a specific time interval.

Medieval Warm Period: Warm period between AD 725 and 1025.

meteorological tides: Water-level fluctuations caused by winds or changes in barometric pressure.

Milankovitch cycles: Climate cycles driven by variations in the Earth's rotation around its axis and its revolution around the sun.

obliquity: The tilt of the Earth's axis of rotation around the sun.

oceanographic tides: Changes in sea-surface elevation controlled by oceanographic circulation.

optically stimulated luminescence (OSL): A method to measure the age of a deposit. Following deposition, quartz grains are exposed to radioactive decay, which varies with burial time. The trapped energy released by exposure to light is controlled by the time of burial and yields a measure of the age of a deposit.

ostracods: Microscopic crustaceans with two valves, ranging from freshwater to deep marine habitats. They are relatively abundant in estuaries and are good indicators of average salinity.

overwash: The transport of sediment across a coastal barrier and its deposition along the back side of the barrier during storms.

Paleocene-Eocene Thermal Maximum: A geological period of unusually warm atmospheric conditions marked by exceptionally high greenhouse gas concentrations.

paleoshoreline: Shoreline position from a previous sea-level location.

paleotempestology: A relatively young research field that uses storm deposits to reconstruct hurricane activity at geological timescales.

Palmer Drought Severity Index (PDSI): A standardized index that uses available temperature and precipitation data to estimate relative dryness, on a scale from −10 (dry) to +10 (wet), and to measure periods of extreme drought.

pressure melting: Meltwater production at the base of an ice sheet resulting from the weight of the ice sheet.

progradation: The seaward movement of a shoreline or delta when sediment supply outpaces relative sea-level rise.

progradational barrier: A barrier island with a history of seaward growth.

recessional moraines: Linear ridges formed by accumulation of sediments at the terminus of a retreating glacier or ice sheet.

regression: Seaward shoreline migration.

relative sea-level rise (RSLR): Net sea-level rise resulting from a combination of eustasy, subsidence, and glacial isostatic effects.

rip current: An offshore-directed coastal current.

river-influenced estuaries: Estuaries formed as sea level flooded river valleys and maintained connectivity to these rivers.

rollover: Landward migration of barrier islands where the rate of washover outpaces sand supply.

salt marsh: A coastal wetland exposed to marine influence during flood tides.

salt tectonics: Deformation resulting from the movement of subsurface salt.

sedimentary facies: A sediment type unique to a specific environment.

sediment budget: Quantitative analysis of sediment inputs and outputs that identifies sediment sources and sinks.

sediment isostatic adjustment (SIA): Subsidence driven by the weight of the sediment column.

seismic facies: Seismic reflection character unique to a specific depositional process or environment.

shoreface (zone): The shallow, nearshore coastal environment dominated by wave motion and circulation.

SLOSH model: A model used by the National Hurricane Center to predict the magnitude of storm surge using multiple factors, including the slope of the continental shelf and coastal plain, the central pressure of the storm, wind speeds, storm size and speed, the angle of storm approach, and astronomical tides.

steric influence: Sea-level rise caused by warming and expansion of oceans.

storm surge: A rapid rise in sea level that results from atmospheric pressure changes and storm winds.

storm surge channel: A channel that cuts though a barrier island and is formed during a storm surge.

surf zone: The nearshore zone characterized by breaking waves.

swash zone: The intertidal part of a beach where wave swash dominates.

swath bathymetry: Technology used to obtain 2-D images of the seafloor using multiple transducers mounted in the hull of a vessel.

tectonic subsidence: Basin-scale subsidence that results from large-scale tectonic forces.

Texas Mud Blanket: A layer of Holocene mud that extends across the central Texas continental shelf.

tidal delta: A sediment body formed by flood tides.

tidal inlet: A natural channel that connects an estuary to the offshore setting.

tipping point: A critical point at which a sudden and dramatic shift to a contrasting dynamic regime occurs.

toe of shoreface: The bathymetric break in the offshore profile that coincides with the boundary between the shoreface and offshore marine environment.

transgression: Landward migration of a shoreline.

transgressive barrier: A barrier island migrating landward.

transgressive ravinement: The process of erosion by storm waves during transgression.

transgressive ravinement surface: The erosion surface formed during transgression.

vertical accretion: The rate of upward growth of wetlands by accumulation of new sediment (both terrigenous and organic).

washover: Sediment transported from a coastal barrier into an estuary or a lagoon during storms.

washover fan: An accumulation of storm washover deposits, commonly associated with a breach in a coastal barrier.

wave-dominated delta: A delta formed when sediment delivery from a river is overshadowed by wave reworking, resulting in a lobate shoreline and restricted offshore delta development.

REFERENCES

Alley, R. B., Blankenship, D. D., Bentley, C. R., and Rooney, S. T. 1986. Deformation of till beneath ice stream B, West Antarctica. *Nature* 322:57–59.

Anderson, J. B., and Rodriguez, A. B., eds. 2008. *Response of Upper Gulf Coast Estuaries to Holocene Climate Change and Sea-Level Rise*. Geological Society of America Special Paper 443. https://doi.org/10.1130/SPE443.

Anderson, J. B., and Thomas, M. A. 1991. Marine ice-sheet decoupling as a mechanism for rapid, episodic sea-level change: The record of such events and their influence on sedimentation. *Sedimentary Geology* 70:87–104. https://doi.org/10.1016/0037-0738(91)90136-2.

Anderson, J. B., Rodriguez, A., Abdulah, K., Fillon, R., Banfield, L. A., McKeown, H., and Wellner, J. 2004. Late Quaternary stratigraphic evolution of the northern Gulf of Mexico margin: A synthesis. In *Late Quaternary Stratigraphic Evolution of the Northern Gulf of Mexico Margin*, edited by J. B. Anderson and R. H. Fillon, Society for Sedimentary Geology Special Publication 79. 1–24. https://doi.org/10.2110/pec.04.79.0001.

Anderson, J. B., Rodriguez, A. B., Milliken, K., and Taviani, M. 2008. The Holocene Evolution of the Galveston Bay complex, Texas: Evidence for rapid change in estuarine environments. In *Response of Upper Gulf Coast Estuaries to Holocene Climate Change and Sea-Level Rise*, edited by J. B. Anderson and A. B. Rodriguez, 89–104. Geological Society of America Special Paper 443.

Anderson, J. B., Conway, H., Bart, P. J., Witus, A. E., Greenwood, S. L., McKay, R. M., Hall, B. L., et al. 2014. Ross Sea paleo-ice sheet drainage and deglacial history during and since the LGM. *Quaternary Science Reviews* 100:31–54.

Anderson, J. B., Wallace, D. J., Rodriguez, A. B., Simms, A. B., and Milliken, K. T. 2022. *Holocene Evolution of the Western Louisiana-Texas Coast, USA: Response to Sea-Level Rise and Climate Change*. Geological Society of America Memoir 221.

Anderson, J. B., Wallace, D. J., Rodriguez, A. B., and Simms, A. R. 2023. Unprecedented historical erosion of US Gulf Coast: A consequence of accelerated sea-level rise? *Earth's Future* 11, e2023EF003676. https://doi.org/10.1029/2023EF003676.

Balaguru, K., Chang, C. C., Leung, L. R., Foltz, G. R., Hagos, S. M., Wehner, M. F., Kossin, J. P., Ting, M., and Xu, W. 2024. A global increase in nearshore tropical cyclone intensification. *Earth's Future* 12, e2023EF004230. https://doi .org/10.1029/2023EF004230.

Balsillie, J. H., and Donoghue, J. F. 2004. *High-Resolution Sea-Level History for the Gulf of Mexico since the Last Glacial Maximum.* Florida Geological Survey Report of Investigations No. 103.

Banfield, L., and Anderson, J. B. 2004. The Late Quaternary evolution of the Rio Grande Delta: Complex response to eustasy and climate change. In *Late Quaternary Stratigraphic Evolution of the Northern Gulf of Mexico Margin,* edited by J. B. Anderson and R. H. Fillon, 289–306. Society for Sedimentary Geology Special Publication 79. https://doi.org/10.2110/pec.04.79.0289.

Bart, P. J., and Anderson, J. B. 2004. Late Quaternary stratigraphic evolution of the Alabama-West Florida outer continental shelf, Gulf of Mexico: A synthesis. In *Late Quaternary Stratigraphic Evolution of the Northern Gulf of Mexico Margin,* edited by J. B. Anderson and R. H. Fillon, 43–54. Society for Sedimentary Geology Special Publication 79.

Bart, P. J., and Kratochvil, M. 2022. A paleo-perspective on West Antarctic Ice Sheet retreat. *Scientific Reports* 21. https://doi.org/10.1038/s41598-022 -22450-3.

Benke, A., and Cushing, C. 2005. *Rivers of North America.* Burlington, MA: Elsevier Academic Press.

Bentley, M. J., O'Cofaigh, C., Anderson, J. B., Conway, H., Davies, B., Graham, A. G. C., Hillenbrand, C. C., et. al. 2014. A community-based geological reconstruction of Antarctic Ice Sheet deglaciation since the Last Glacial Maximum. *Quaternary Science Reviews* 100:1–9. https://doi.org/10.1016/j .quascirev.2014.06.025.

Blum, M. D., and Roberts, H. H. 2009. Drowning of the Mississippi Delta due to insufficient sediment supply and global sea-level rise. *Nature Geoscience* 2:488–91.

Bregy, J. C., Wallace, D. J., Totten, R., and Cruz, V. J. 2018. 2500-year paleotempestological record of intense storms for the northern Gulf of Mexico, United States. *Marine Geology* 396:26–42. https://doi.org/10.1016 /j.margeo.2017.09.009.

Bregy, J. C., Maxwell, J. T., Robeson, S. M., Harley, G. L., Elliott, E. A., and Heeter, K. J. 2022. US Gulf Coast tropical cyclone precipitation influenced by volcanism and the North Atlantic subtropical high. *Communications Earth & Environment* 3. https://doi.org/10.1038/s43247-022-00494-7.

Bruyère, C. L., Buckley, B., Jaye, A. B., Done, J. M., Leplastrier, M., Aldridge, J., Chan, P., Towler, E., and Ge, M. 2022. Using large climate model ensembles to assess historical and future tropical cyclone activity along the Australian east coast. *Weather and Climate Extremes* 38. https://doi.org/10.1016/j.wace .2022.100507.

Burstein, J. T., Goff, J. A., Gulick, S. P. S., Lowery, C., Standring, P., and Swartz, J. 2023. Tracking barrier island response to early Holocene sea-level rise: High resolution study of estuarine sediments in the Trinity River Paleovalley. *Marine Geology* 455. https://doi.org/10.1016/j.margeo.2022.106951.

Buzas-Stephens, P., Livsey, D. N., Simms, A. R., and Buzas, M. A. 2014. Estuarine foraminifera record Holocene stratigraphic changes and Holocene climate changes in ENSO and the North American monsoon: Baffin Bay, Texas. *Palaeogeography, Palaeoclimatology, Palaeoecology* 404:44–56. https://doi.org /10.1016/j.palaeo.2014.03.031.

Byrnes, M. R., Britsch, L. D., Berlinghoff, J. L., Johnson, R., and Khalil, S. 2019. Recent subsidence rates for Barataria Basin, Louisiana. *Geo-Marine Letters* 39:265–78, https://doi.org/10.1007/s00367-019-00573-3.

Capistrant-Fossa, K. A., and Dunton, K. H. 2024. Rapid sea level rise causes loss of seagrass meadows. *Communications Earth & Environment* 5. https://doi .org/10.1038/s43247-024-01236-7.

Costanza, R., de Groot, R., Sutton, P., van der Ploeg, S., Anderson, S., Kubiszewski, I., Farber, S., and Turner, R. K. 2014. Changes in the global value of ecosystem services. *Global Environmental Change* 26:152–58.

Couvillion, B. R., Beck, H., Schoolmaster, D., and Fischer, M. 2017. Land area change in coastal Louisiana (1932 to 2016). US Geological Survey Scientific Investigations Map 3381. https://doi.org/10.3133/sim3381.

Dangendorf, S., Hendricks, N., Qiang, S., Klink, J., Ezer, T., Frederikse, T., Calafat, F. M., Wahl, T., and Törnqvist, T. E. 2023. Acceleration of U.S. Southeast and Gulf Coast sea-level rise amplified by internal climate variability. *Nature Communications* 14. https://doi.org/10.1038/s41467-023 -37649-9.

Davis, R. A., Jr., Elko, N., and Wang, P. 2018. *Managing the Gulf Coast Using Geology and Engineering.* Geological Society of America. https://doi. org/10.1130/9780813741239.

Day, J. W., Jr., Boesch, D. F., Clairain, E. J., Kemp, G. P., Laska, S. B., Mitsch, W. J., Orth, K., et al. 2007. Restoration of the Mississippi Delta: Lessons from hurricanes Katrina and Rita. *Science* 315:1679–84.

Dean, R. G., and Houston, J. R. 2016. Determining shoreline response to sea level rise. *Coastal Engineering* 114:1–8.

Dellapenna, T., Hoelscher, C., Hill, L., Mukaimi, M., and Knap, M. 2020. How tropical cyclone flooding caused erosion and dispersal of mercury-contaminated sediment in an urban estuary: The impact of Hurricane Harvey on Buffalo Bayou and the San Jacinto Estuary, Galveston Bay, USA. *Science of the Total Environment* 748. https://doi.org/10.1016/j.scitotenv.2020.141226.

DeLong, K. L., Gonzalez, S., Obelcz, J. B., Truong, J. T., Bentley, S. J., Xu, K., Reese, C. A., et al. 2021. Late Pleistocene baldcypress (*Taxodium distichum*) forest deposit on the continental shelf of the northern Gulf of Mexico. *Boreas* 50:871–92. https://onlinelibrary.wiley.com/doi/full/10.1111/bor.12524.

Duke, T., and Kruczynski, W. L. 1992. *Report on the Status and Trends of Emergent and Submerged Vegetated Habitats of Gulf of Mexico Coastal Waters, U.S.A.* US Environmental Protection Agency 800-R-92-00.

Dunion, J. P., and Velden, C. S. 2004. The impact of the Saharan Air Layer on Atlantic tropical cyclone activity. *Bulletin of the American Meteorological Society* 85:353–65.

Emanuel, K. 2005. Increasing destructiveness of tropical cyclones over the past 30 years. *Nature* 436:686–88.

Ferguson, S., Warny, S., Anderson, J. B., Simms, A. R., and White, C. 2018a. Breaching of Mustang Island in response to the 8.2 ka sea-level event and impact on Corpus Christi Bay, Gulf of Mexico: Implications for future coastal change. *The Holocene* 28:166–72. https://doi.org/10.1177/0959683617715697.

Ferguson, S., Warny, S., Anderson, J. B., Simms, A. R., and Escarguel, G. 2018b. Holocene vegetation and climate evolution of Corpus Christi and Trinity bays: Implications on source-to-sink deposition on the Texas coast. *Geo-Bios* 51:123–35. https://doi.org/10.1016/j.geobios.2018.02.007.

Fraticelli, C. M. 2006. Climate forcing in a wave-dominated delta: The effects of drought-flood cycles on delta progradation. *Journal of Sedimentary Research* 76:1067–76. https://doi.org/10.2110/jsr.2006.097.

Gabrysch, R. K., and Bonnet, C. W. 1975. *Land-Surface Subsidence at Seabrook, Texas.* US Geological Survey Open-File Report 75-413. https://doi:10.3133/ofr75413.

Gerlach, M. J., Engelhart, S. E., Kemp, C. A., Moyer, R. P., Smoak, J. M., Bernhardt, E., and Cahill, N. 2017. Reconstructing Common Era relative sea-level change on the Gulf Coast of Florida. *Marine Geology* 390:254–69. https://doi.org/10.1016/j.margeo.2017.07.001.

Goff, J. A., Lugrin, L., Gulick, S. P., Thirumalai, K., and Okumura, Y. 2016. Oyster reef die-offs in stratigraphic record of Corpus Christi Bay, Texas, possibly caused by drought-driven extreme salinity changes. *The Holocene* 26:511–19. https://doi.org/10.1177/0959683615612587.

Graham, A. G. C., Wåhlin, A., Hogan, K. A., Nitsche, F. O., Heywood, K. J., Minzoni, R., Smith, J. A., et al. 2022. Rapid retreat of Thwaites Glacier from a pinning point in the pre-satellite era. *Nature Geoscience* 15:687–88.

Griggs, G. 2017. *Coasts in Crisis: A Global Challenge.* Berkeley: University of California Press.

Halligan, J. J., Waters, M. R., Perrotti, A., Owens, I. J., Feinberg, J. M., Bourne, M. D., Fenerty, B., et al. 2016. Pre-Clovis occupation 14,550 years ago at the Page-Ladson Site, Florida and the peopling of the Americas. *Science Advances* 2, e1600375. https://doi:10.1126/sciadv.1600375.

Hijma, M. P., Shen, Z., Torbjörn, T., and Mauz, B. E. 2017. Late Holocene evolution of a coupled, mud-dominated delta plain–chenier plain system, coastal Louisiana, USA. *Earth Surface Dynamics* 5:689–710. https://doi.org/10.5194/esurf-5-689-2017.

Himmelstoss, E. A., Kratzmann, M. G., and Thieler, E. R. 2017. National assessment of shoreline change: A GIS compilation of updated vector shorelines and associated shoreline change data for the Gulf of Mexico. US Geological Survey data release. https://doi.org/10.5066/F78P5XNK.

Hine, A. C. 2013. *Geologic History of Florida*. Gainesville: University Press of Florida.

Hine, A. C., Chambers, D. P., Clayton, T. D., Hafen, M. R., and Mitchum, G. T. 2016. *Sea Level Rise in Florida: Science, Impacts, and Options*. Gainesville: University Press of Florida.

Hollis, R. J., Wallace, D. J., Miner, M. D., Gal, N. S., Dike, C., and Flocks, J. G. 2019. Late Quaternary evolution and stratigraphic framework influence on coastal systems along the north-central Gulf of Mexico, USA. *Quaternary Science Reviews* 223, 105910.

Hörhold, M., Münch, T., Weißbach, S., Kipfstuhl, S., Freitag, J., Sasgen, I., Lohmann, G., Vinther, B., and Laepple, T. 2023. Modern temperatures in central–north Greenland warmest in past millennium. *Nature* 613:503–7. https://doi.org/10.1038/s41586-022-05517-z.

Hughes, T. 1973. Is the West Antarctic Ice Sheet disintegrating? *Journal of Geophysical Research* 78:7884–910.

Jakobsson, M., Anderson, J. B., Nitsche, F., Dowdeswell, J. A., Gyllencreutz, R., Kirchner, N., Mohammad, R., et al. 2011. Geological record of ice shelf break-up and grounding line retreat, Pine Island Bay, West Antarctica. *Geology* 39:691–94.

Jankowski, K. L., Törnqvist, T. E., and Fernandes, A. M. 2017. Vulnerability of Louisiana's coastal wetlands to present-day rates of relative sea-level rise. *Nature Communications* 8, 14792. https://doi.org/10.1038/ncomms14792.

Kemp, A. C., Horton, B. P., Culver, S. J., Corbett, D. R., van de Plassche, O., Gehrels, W. R., Douglas, B. C., and Parnell, A. C. 2009. Timing and magnitude of recent accelerated sea-level rise (North Carolina, United States). *Geology* 37:1035–38. https://doi.org/10.1130/G30352A.1.

Kirwan, M. L., Temmerman, S., Skeehan, E. E., Guntenspergen, G. R., and Fagherazzi, S. 2016. Overestimation of marsh vulnerability to sea level rise. *Nature Climate Change* 6:253–60.

Knutson, T., Camargo, S. J., Chan, J. C. L., Emanuel, K., Ho, C. H., Kossin, J., Mohapatra, M., et al. 2019. Tropical cyclones and climate change assessment, Part 1: detection and attribution. *Bulletin of the American Meteorological Society* 100(10):1987–2007. https://doi.org/10.1175/BAMS-D-18-0189.1.

Lane, P., Donnelly, J. P., Woodruff, J. D., and Hawkes, A. D. 2011. A decadally-resolved paleo-hurricane record archived in the late Holocene sediments of a Florida sinkhole. *Marine Geology* 287:14–30.

Li, G., Törnqvist, T. E., and Dangendorf, S. 2024. Real-world time-travel experiment shows ecosystem collapse due to anthropogenic climate

change. *Nature Communications* 15, 1226. https://doi.org/10.1038/s41467
  -024-45487-6.

Li, Y.-X., Törnqvist, T. E., Nevitt, M., and Kohl, B. 2012. Synchronizing a sea-level
  jump, final Lake Agassiz drainage, and abrupt cooling 8200 years ago. *Earth
  and Planetary Science Letters* 315–16:41–50.

Liénard, J., Harrison, J., and Strigul, N. 2016. US forest response to projected
  climate-related stress: A tolerance perspective. *Global Change Biology*
  22:2875–86. https://doi.org/10.1111/gcb.13291.

Liu, K., and Fearn, M. L. 2000a. Holocene history of catastrophic hurricane
  landfalls along the Gulf of Mexico coast reconstructed from coastal lake and
  marsh sediments. In *Current Stresses and Potential Vulnerabilities: Implications
  of Global Change for the Gulf Coast Region of the United States*, edited by Z. H.
  Ning and K. K. Abdollahi, 38–47. Baton Rouge: Franklin Press.

Liu, K., and Fearn, M. L. 2000b. Reconstruction of prehistoric landfall
  frequencies of catastrophic hurricanes in northwestern Florida from lake
  sediment records. *Quaternary Research* 54:238–45. https://doi:10.1006
  /qres.2000.2166.

Livsey, D., and Simms, A. R. 2013. Holocene sea-level change derived from
  microbial mats. *Geology* 41:971–74.

Livsey, D., and Simms, A. R. 2016. Episodic flooding of estuarine environments
  in response to drying climate over the last 6,000 years in Baffin Bay, Texas.
  Marine Geology 381:142–62. https://doi.org/10.1016/j.margeo.2016.09.003.

López, G. I., and Rink, W. J. 2008. New quartz optical stimulated luminescence
  ages for beach ridges on the St. Vincent Island Holocene strandplain, Florida,
  United States. *Journal of Coastal Research* 24:49–62. https://doi.org/10.2112
  /05-0473.1.

Lorenzo-Trueba, J., and Ashton, A. D. 2014. Rollover, drowning, and
  discontinuous retreat: Distinct modes of barrier response to sea-level rise
  arising from a simple morphodynamic model. *Journal of Geophysical Research:
  Earth Surface* 119:779–801. https://doi.org/10.1002/2013JF002941.

Maddox, J., Anderson, J. B., Milliken, K., and Rodriguez, A. B., Dellapenna,
  T. M, and Giosan, L. 2008. The Holocene evolution of the Matagorda and
  Lavaca estuary complex, Texas, USA. In *Response of Upper Gulf Coast
  Estuaries to Holocene Climate Change and Sea-Level Rise*, edited by J. B.
  Anderson and A. B. Rodriguez, 105–19. Geological Society of America S
  pecial Paper 443. https://doi.org/10.1130/2008.2443(07).

Mann, M. E., Bradley, R. S., and Hughes, M. K. 1998. Global-scale temperature
  patterns and climate forcing over the past six centuries. *Nature* 392:779–87.

Mann, M. E., Peterson, T. C., and Hassol, S. J. 2017. What we know about the
  climate change–hurricane connection. *Environmental Science.*
  https://api.semanticscholar.org/CorpusID:248096805.

Marcott, S., Shakun, J., Clark, P., and Mix, A. 2013. A reconstruction of regional
  and global temperature for the past 11,300 years. *Science* 339:198–201. https://
  doi.org/10.1126/science.1228026.

Mariotti, G., and Hein, C. J., 2022. Lag in response of coastal barrier-island retreat to sea-level rise. *Nature Geoscience* 15:633–38. https://doi.org/10.1038/s41561-022-00980-9.

McBride, R. A., Moslow, T. F., Roberts, H. H., and Diecchio, R. J. 2004. Late Quaternary geology of the northeastern Gulf of Mexico shelf: Sedimentology, depositional history, and ancient analogs of a major shelf sand of the modern transgressive system. In *Late Quaternary Stratigraphic Evolution of the Northern Gulf of Mexico Margin*, edited by J. B. Anderson and R. H. Fillon, 1–24. Society for Sedimentary Geology Special Publication 79. https://doi.org/10.2110/pec.04.79.0001.

McBride, R. A., Taylor, M. J., and Byrnes, M. R., 2007. Coastal morphodynamics and Chenier-Plain evolution in southwestern Louisiana, USA: A geomorphic model. *Geomorphology* 88:367–422. https://doi.org/10.1016/j.geomorph.2006.11.013.

McBride, R. A., Anderson, J. B., Buynevich, I. V., Byrnes, M. R., Cleary, W., Fenster, M. S., FitzGerald, D. M., et al. 2022. Morphodynamics of modern and ancient barrier systems: An updated and expanded synthesis. In *Treatise on Geomorphology*, vol. 8, edited by J. J. F. Shroder, 289–417. Elsevier, Academic Press. https://dx.doi.org/10.1016/B978-0-12-818234-5.00153-X.

McKeown, H., Bart, P., and Anderson, J. B. 2004. High-resolution stratigraphy of a sandy, ramp-type margin: Offshore Apalachicola, Florida. In *Late Quaternary Stratigraphic Evolution of the Northern Gulf of Mexico Margin*, edited by J. B. Anderson and R. H. Fillon, 25–42. Society for Sedimentary Geology Special Publication 79.

Mercer, J. H. 1978. West Antarctic ice sheet and CO_2 greenhouse effect: A threat of disaster. *Nature* 271:321–25.

Millan, R., Mouginot, J., Rabatel, A., and Morlighem, M. 2022. Ice velocity and thickness of the world's glaciers. *Nature Geoscience* 15:124–29. https://doi.org/10.1038/s41561-021-00885-z.

Milliken, K., Anderson, J. B., Rodriguez, A. B., and Simms, A. 2008a. A new composite Holocene sea-level curve for the northern Gulf of Mexico. In *Response of Upper Gulf Coast Estuaries to Holocene Climate Change and Sea-Level Rise*, edited by J. B. Anderson and A. B. Rodriguez, 1–11. Geological Society of America Special Paper 443.

Milliken, K. T., Anderson, J. B., and Rodriguez, A. B. 2008b. Record of dramatic Holocene environmental changes linked to eustasy and climate change in Calcasieu Lake, Louisiana, USA. In *Response of Upper Gulf Coast Estuaries to Holocene Climate Change and Sea-Level Rise*, edited by J. B. Anderson and A. B. Rodriguez, 43–63. Geological Society of America Special Paper 443. https://doi.org/10.1130/2008.2443(04).

Milliken, K. T., Anderson, J. B., and Rodriguez, A. B. 2008c. Tracking the Holocene evolution of Sabine Lake through the interplay of eustasy, antecedent topography, and sediment supply variations, Texas and Louisiana,

USA. In *Response of Upper Gulf Coast Estuaries to Holocene Climate Change and Sea-Level Rise*, edited by J. B. Anderson and A. B. Rodriguez, 65–88. Geological Society of America Special Paper 443. https://doi.org/10.1130 /2008.2443(05).

Milliken, K. T., Anderson, J. B., Simms, A., and Blum, M. D. 2017. A Holocene record of flux of alluvial sediment related to climate: Case studies from the northern Gulf of Mexico. *Journal of Sedimentary Research* 87:780–94. https:// doi.org/10.2110/jsr.2017.43.

Mitrovica, J. X., Gomez, N., and Clark, P. U. 2009. The sea-level fingerprint of West Antarctic collapse. *Science* 323:753.

Monica, S. B., Wallace, D. J., Wallace, E. J., Du, X, Dee, S. G., and Anderson, J. B. 2024. 4500-year paleohurricane record from the western Gulf of Mexico, Coastal Central Texas. *Marine Geology* 473. https://doi.org/10.1016/j.margeo .2024.107303.

Morris, J. T., Sundareshwar, P. V., Nietch, C. T., Kjerfve, B., and Cahoon, D. R. 2002. Responses of coastal wetlands to rising sea level. *Ecology* 83:2869–77.

Morton, R. A. 2007. *Historical Changes in the Mississippi-Alabama Barrier Islands and the Roles of Extreme Storms, Sea Level, and Human Activities*. US Geological Survey Open-File Report 2007-1161.

Morton, R. A., Buster, N. A., and Krohn, M. D. 2002. Subsurface controls on historical subsidence rates and associated wetland loss in southcentral Louisiana. *Transactions of the Gulf Coast Association of Geological Societies* 52:767–78.

Morton, R. A., Miller, T. L., and Moore, L. J. 2004. *National Assessment of Shoreline Change: Part 1, Historical Shoreline Changes and Associated Coastal Land Loss along the U.S. Gulf of Mexico*. US Geological Survey Open-File Report 2004–1043.

Nielsen-Gammon, J. W., Banner, J. L., Cook, B. I., Tremaine, D. M., Wong, C. I., Mace, R. E., Gao, H., et al. 2020. Unprecedented drought challenges for Texas water resources in a changing climate: What do researchers and stakeholders need to know? *Earth's Future* 8, e2020EF001552. https://doi.org/10.1029 /2020EF001552.

Nielsen-Gammon, J., Holman, S., Buley, A., Jorgensen, S., Escobedo, J., Ott, C., Dedrick, J., and Van Fleet, A. 2021. *Assessment of Historic and Future Trends of Extreme Weather in Texas, 1900–2036: 2021 Update*. Document OSC-202101. College Station: Texas A&M University, Office of the State Climatologist.

Nienhuis, J. H., and Lorenzo-Trueba, J. 2019. Can barrier islands survive sea-level rise? Quantifying the relative role of tidal inlets and overwash deposi-tion. *Geophysical Research Letters* 46:14613–21. https://doi.org/10.1029 /2019GL085524.

Odezulu, C. I., Lorenzo-Trueba, J., Wallace, D. J., and Anderson, J. B. 2018. Follets Island: A case of unprecedented change and transition from rollover to subaqueous shoals. In *Barrier Dynamics and Response to Changing Climate*, edited by L. J. Moore and A. B. Murray, 147–74. Cham, Switzerland: Springer. https://doi.org/10.1007/978-3-319-68086-6_5.

Odezulu, C. I., Swanson, T., and Anderson, J. B. 2020. Holocene progradation and retrogradation of the central Texas coast regulated by alongshore and cross-shore sediment flux variability. *The Depositional Record* 7:77–92. https://doi.org/10.1002/dep2.130.

Ohenhen, L. O., Shirzaei, M., Ojha, C., and Kirwan, M. L. 2023. Hidden vulnerability of US Atlantic coast to sea-level rise due to vertical land motion. *Nature Communications* 14, 2038. https://doi.org/10.1038/s41467-023-37853-7.

Osland, M. J., Chivoiu, B., Enwright, N. M., Thorne, K. M., Guntenspergen, G. R., Grace, J. B., Dale, L. L., et al. 2022. Migration and transformation of coastal wetlands in response to rising seas. *Science Advances* 8. https://doi.org/10.1126/sciadv.abo5174.

Osterman, L. E., Twichell, D. C., and Poore, R. Z. 2009. Holocene evolution of Apalachicola Bay, Florida. *Geo-Marine Letters* 29:395–404.

Otvos, E. G. 1979. Barrier island evolution and history of migration, north central Gulf Coast. In *Barrier Islands*, edited by S. P. Leatherman, 291–319. New York: Academic Press.

Otvos, E. G. 2018. Coastal barriers, northern Gulf: Last eustatic cycle; genetic categories and development contrasts. A review. *Quaternary Science Reviews* 193:212–43. https://doi.org/10.1016/j.quascirev.2018.04.001.

Otvos, E. G., and Giardino, M. J. 2004. Interlinked barrier chain and delta lobe development, northern Gulf of Mexico. *Journal of Coastal Research* 242: 463–78.

Paine, J. G., Mathew, S., and Caudle, T. 2014. Historical shoreline change through 2007, Texas Gulf Coast: Rates, contributing causes, and Holocene context. *Gulf Coast Association of Geological Societies Journal* 1:13–26.

Paine, J. G., Caudle, T. L., and Andrews, J. R. 2021. *Shoreline Movement and Beach and Dune Volumetrics along the Texas Gulf Coast, 1930s to 2019.* Texas General Land Office Final Report, CEPRA Project 1662.

Parkinson, R. W., and Wdowinski, S. 2023. A unified conceptual model of coastal response to accelerating sea level rise, Florida, U.S.A. *Science of the Total Environment* 892, 164448.

Pfleiderer, P., Nath, S., and Schleussner, C.-F. 2022. Extreme Atlantic hurricane seasons made twice as likely by ocean warming. *Weather and Climate Dynamics* 3:471–82.

Phillips, J. D. 2010. Relative importance of intrinsic, extrinsic, and anthropic factors in the geomorphic zonation of the Trinity River, Texas. *Journal of the American Water Resources Association* 46:807–23. https://doi.org/10.1111/j.1093-474X.2010.00457.x.

Piecuch, C. G. 2020. Likely weakening of the Florida Current during the past century revealed by sea-level observations. *Nature Communications* 11, 3973. https://doi.org/10.1038/s41467-020-17761-w.

Prothro, L. O., Majewski, W., Yokoyama, Y., Simkins, L. M., Anderson, J. B., Yamane, M., Miyairi, Y., and Ohkouchi, N. 2020. Timing and pathways of East Antarctic Ice Sheet retreat. *Quaternary Science Reviews* 230, 106166. https://doi.org/10.1016/j.quascirev.2020.106166.

Rasmussen, S. O., Andersen, K. K., Svensson, A. M., Steffensen, J. P., Vinther, B. M., Clausen, H. B., Siggaard-Andersen, M. L., et al. 2006. A new Greenland ice core chronology for the last glacial termination. *Journal of Geophysical Research: Atmospheres* 111, D06102. https://doi.org/10.1029/2005JD006079.

Rice, J. A., Simms, A. R., Buzas-Stephens, P., Steel, E., Livsey, D., Reynolds, L. C., Yokoyama, Y., and Halihan, T. 2020. Deltaic response to climate change: The Holocene history of the Nueces Delta. *Global and Planetary Change* 191. https://doi.org/10.1016/j.gloplacha.2020.103213.

Rignot, E. 2021. Sea level rise from melting glaciers and ice sheets caused by climate warming above pre-industrial levels. *Physics-Uspekhi* 65:1129–38. https://doi.org/10.3367/UFNe.2021.11.039106.

Rink, W. J., and López, G. I. 2010. OSL-based lateral progradation and aeolian sediment accumulation rates for the Apalachicola barrier island complex, north Gulf of Mexico, Florida. *Geomorphology* 123:330–42.

Rodriguez, A. B., and Meyer, C. T. 2006. Sea level variation during the Holocene deduced from the morphologic and stratigraphic evolution of Morgan Peninsula, Alabama, USA. *Journal of Sedimentary Research* 76:257–69.

Rodriguez, A. B., Anderson, J. B., Siringan, F. P., and Taviani, M. 1999. Sedimentary facies and genesis of Holocene sand banks on the east Texas inner continental shelf. In *Isolated Shallow Marine Sand Bodies*, edited by J. Sneddon and K. Bergman, 165–78. Society for Sedimentary Geology Special Publication 64. https://doi.org/10.2110/pec.99.64.0165.

Rodriguez, A. B., Hamilton, M., and Anderson, J. B. 2000. Facies and evolution of the modern Brazos Delta, Texas: Wave versus flood influence. *Journal of Sedimentary Research* 70:283–95. https://doi.org/10.1306/2DC40911-0E47-11D7-8643000102C1865D.

Rodriguez, A. B., Fassell, M., and Anderson, J. B. 2001. Variations in shoreface progradation and ravinement along the Texas coast, Gulf of Mexico. *Sedimentology* 48:837–53. https://doi.org/10.1046/j.1365-3091.2001.00390.x.

Rodriguez, A. B., Anderson, J. B., Siringan, F. P., and Taviani, M. 2004. Holocene evolution of the east Texas Coast and inner continental shelf: Along-strike variability in coastal retreat rates. *Journal of Sedimentary Research* 74:404–21. https://doi.org/10.1306/092403740405.

Rodriguez, A. B., Greene, L. D., Anderson, J. B., and Simms, A. R. 2008. Response of Mobile Bay and eastern Mississippi Sound, Alabama to changes in sediment accommodation and accumulation. In *Response of Upper Gulf Coast Estuaries to Holocene Climate Change and Sea-Level Rise*, edited by J. B. Anderson and A. B. Rodriguez, 13–29. Geological Society of America Special Paper 443.

Rodriguez, A. B., Simms, A. R., and Anderson, J. B. 2010. Bay-head deltas across the northern Gulf of Mexico back step in response to the 8.2 ka cooling event. *Quaternary Science Reviews* 29:3983–93. https://doi.org/10.1016/j.quascirev.2010.10.004.

Rodysill, J. R., Donnelly, J. P., Sullivan, R., Lane, P. D., Toomey, M., Woodruff, J. D., Hawkes, A. D., et al. 2020. Historically unprecedented northern Gulf of Mexico hurricane activity from 650 to 1250 CE. *Scientific Reports* 10, 19092. https://doi.org/10.1038/s41598-020-75874-0.

Saintilan, N., Khan, N. S., Ashe, E., Kelleway, J. J., Rogers, K., Woodroffe, C. D., and Horton, B. P. 2020. Thresholds of mangrove survival under rapid sea level rise. *Science* 368:1118–21.

Schuerch, M., Spencer, T., Temmerman, S., Kirwan, M. L., Wolff, C., Lincke, D., McOwen, C. J., et al. 2018. Future response of global coastal wetlands to sea-level rise. *Nature* 561:231–34. https://doi.org/10.1038/s41586-018-0476-5.

Seavey, J. R., Pine, W. E., III, Frederick, P., Sturmer, L., and Berrigan, M. 2011. Decadal changes in oyster reefs in the Big Bend of Florida's Gulf Coast. *Ecosphere* 2:1–14. https://doi.org/10.1890/ES11-00205.1.

Shinkle, K. D., and Dokka, R. K. 2004. *Rates of Vertical Displacement at Benchmarks in the Lower Mississippi Valley and the Northern Gulf Coast.* NOAA Technical Report 50.

Simms, A. R., Anderson, J. B., and Blum, M. 2006. Barrier-island aggradation via inlet migration: Mustang Island, Texas. *Sedimentary Geology* 187:105–25. https://doi.org/10.1016/j.sedgeo.2005.12.023.

Simms, A. R., Anderson, J. B., Milliken, K. T., Taha, Z. P., and Wellner, J. S. 2007a. Geomorphology and age of the oxygen isotope stage 2 (last lowstand) sequence boundary on the northwestern Gulf of Mexico continental shelf. In *Seismic Geomorphology: Applications to Hydrocarbon Exploration and Production*, edited by R. J. Davies, H. W. Posamentier, L. J. Wood, and J. A. Cartwright, 29–46. Geological Society, London, Special Publication 277. https://doi.org/10.1144/GSL.SP.2007.277.01.03.

Simms, A. R., Lambeck, K., Purcell, A., Anderson, J., and Rodriguez, A. 2007b. Sea-level history for the Gulf of Mexico since the Last Glacial Maximum with implications for the melting of the Laurentide Ice Sheet. *Quaternary Science Reviews* 26:920–40. https://doi.org/10.1016/j.quascirev.2007.01.001.

Simms, A. R., Anderson, J. B., Rodriguez, A. B., and Taviani, M. 2008. Mechanisms controlling environmental change within an estuary: Corpus Christi Bay, Texas, USA. In *Response of Upper Gulf Coast Estuaries to Holocene Climate Change and Sea-Level Rise*, edited by J. B. Anderson and A. B. Rodriguez, 121–46. Geological Society of America Special Paper 443. https://doi.org/10.1130/2008.2443(08).

Simms, A. R., Aryal, N., Miller, L., and Yokoyama, Y. 2010. The incised valley of Baffin Bay, Texas: A tale of two climates. *Sedimentology* 57:642–69. https://doi.org/10.1111/j.1365-3091.2009.01111.x.

Stapor, F. W., Jr., Mathews, T. D., and Lindfors-Kearns, F. E. 1991. Barrier island progradation and Holocene sea-level history in southwest Florida. *Journal of Coastal Research* 7:815–38.

Sweet, W., Dusek, G. P., Marcy, D. C., Carbin, G. W., and Marra, J. 2019. *2018 State of U.S. High Tide Flooding with a 2019 Outlook*. NOAA Technical Report NOS CO-OPS; 090. https://doi.org/10.25923/rbv9-th19.

Syvitski, J. P. M., Kettner, A. J., Overeem, I., Hutton, E. W. H., Hannon, M. T., Brakenridge, G. R., Day, J., et al. 2009. Sinking deltas due to human activities. *Nature Geoscience* 2:681–86. https://doi.org/10.1038/ngeo629.

Taha, P. Z., and Anderson, J. B. 2008. The influence of valley aggradation and listric normal faulting on styles of river avulsion: A case study of the Brazos River, Texas, USA. *Geomorphology* 95:429–48.

Thomas, M. A., and Anderson, J. B. 1989. Glacial eustatic controls on seismic sequences and parasequences of the Trinity/Sabine incised valley, Texas Continental Shelf. *Transactions of the Gulf Coast Association of Geological Societies* 39:563–70.

Thomas, R. H., and Bentley, C. R. 1978. A model for Holocene retreat of the West Antarctic ice sheet. *Quaternary Research* 10:150–70.

Törnqvist, T. E., and Hijma, M. P. 2012. Links between early Holocene ice-sheet decay, sea-level rise and abrupt climate change. *Nature Geoscience* 5:601–6. https://doi.org/10.1038/ngeo1536.

Törnqvist, T. E., Newsom, L. A., van der Borg, K., de Jong, A. F. M., and Kurnik, C. W. 2004a. Deciphering Holocene sea-level history on the U.S. Gulf Coast: A high-resolution record from the Mississippi Delta. *Geological Society of America Bulletin* 116:1026–39.

Törnqvist, T. E., Bick, S. J., González, J. L., van der Borg, K., and de Jong, A. F. M. 2004b. Tracking the sea-level signature of the 8.2 ka cooling event: New constraints from the Mississippi Delta: *Geophysical Research Letters* 31, L23309. https://doi.org/10.1029/2004GL021429.

Törnqvist, T. E., Bick, S. J., van der Borg, K., and de Jong, A. F. M. 2006. How stable is the Mississippi Delta? *Geology* 34:697–700. https://doi.org/10.1130/G22624.1.

Törnqvist, T. E., Wallace, D. J., Storms, J. E. A., Wallinga, J., van Dam, R. L., Blaauw, M., Derksen, M. S., et al. 2008. Mississippi Delta subsidence primarily caused by compaction of Holocene strata. *Nature Geosciences* 1:173–76.

Törnqvist, T. E., Jankowski, K. L., Li, Y. X., and González, J. L. 2020. Tipping points of Mississippi Delta marshes due to accelerated sea-level rise. *Science Advances* 6, 5512.

Törnqvist, T. E., Cahoon, D. R., Morris, J. T., and Day, J. W. 2021. Coastal wetland resilience, accelerated sea-level rise, and the importance of timescale. *AGU Advances* 2, e2020AV000334. https://doi.org/10.1029/2020AV000334.

Toscano, M. A., and Macintyre, I. G. 2003. Corrected western Atlantic sea-level curve for the last 11,000 years based on calibrated ^{14}C dates from *Acropora palmata* framework and intertidal mangrove peat. *Coral Reefs* 22:257–70. https://doi.org/10.1007/s00338-003-0315-4.

Wahl, E. R., Zorita, E., Diaz, H. F., and Hoell, A. 2022. Southwestern United States drought of the 21st century presages drier conditions into the future. *Communications Earth & Environment* 3, 202. https://doi.org/10.1038/s43247-022-00532-4.

Wallace, D. J., and Anderson, J. B. 2010. Evidence of similar probability of intense hurricane strikes for the Gulf of Mexico over the late Holocene. *Geology* 38:511–14. https://doi.org/10.1130/G30729.1.

Wallace, D. J., and Anderson, J. B. 2013. Unprecedented erosion of the upper Texas coast: Response to accelerated sea-level rise and hurricane impacts. *Geological Society of America Bulletin* 125:728–40. https://doi.org/10.1130/B30725.1.

Wallace, D. J., and Woodruff, J. D. 2020. Experimental and numerical models of fine sediment transport by tsunamis. In *Geological Records of Tsunamis and Other Extreme Waves*, edited by M. Engel, J. Pilarczyk, S. M. May, D. Brill, and E. Garrett, 491–509. http://doi.org/10.1016/B978-0-12-815686-5.00023-7.

Wallace, D. J., Woodruff, J. D., Anderson, J. B., and Donnelly, J. P. 2014. Palaeohurricane reconstructions from sedimentary archives along the Gulf of Mexico, Caribbean Sea, and western North Atlantic Ocean margins. In *Sedimentary Coastal Zones from High to Low Latitudes: Similarities and Differences*, edited by I. P. Martini and H. R. Wanless, 481–501. Geological Society, London, Special Publication 388. https://doi.org/10.1144/SP388.12.

Weight, R., Anderson, J. B., and Fernandez, R. 2011. Rapid mud accumulation on the central Texas shelf linked to climate change and sea-level rise. *Journal of Sedimentary Research* 81:743–64. https://doi.org/10.2110/jsr.2011.57.

Weisberg, R. H., and Zheng, L. 2006. Circulation of Tampa Bay driven by buoyancy, tides, and winds, as simulated using a finite volume coastal ocean model. *Journal Geophysical Research: Oceans* 111. https://doi.org/10.1029/2005JC003067.

White, W. A., Morton, R. A., and Holmes, C. W. 2002. A comparison of factors controlling sedimentation rates and wetland loss in fluvial-deltaic systems, Texas Gulf Coast. *Geomorphology* 44:47–66. https://doi.org/10.1016/S0169-555X(01)00140-4.

Willard, D. A., and Bernhardt, C. E. 2011. Impacts of past climate and sea level change on Everglades wetlands: Placing a century of anthropogenic change into a late-Holocene context. *Climatic Change* 107:59–80. https://doi.org/10.1007/s10584-011-0078-9.

Williams, A. P., Cook, B. I., and Smerdon, J. E. 2022. Rapid intensification of the emerging southwestern North American megadrought in 2020–2021. *Nature Climate Change* 12:232–34. https://doi.org/10.1038/s41558-022-01290-z.

Woodhouse, C. A., Stahle, D. W., and Diaz, J. V. 2012. Rio Grande and Rio Conchos water variability over the past 500 years. *Climate Research* 51:125–36.

Worrall, D. M. 2021. *The Prehistory of Houston and Southeast Texas.* Fulshear, TX: Concertina Press.

Yao, Q., Liu, K., Aragón-Moreno, A. A., Rodrigues, E., Xu, Y. J., and Lam, N. S. 2020. A 5200-year paleoecological and geochemical record of coastal environmental changes and shoreline fluctuations in southwestern Louisiana: Implications for coastal sustainability. *Geomorphology* 365. https://doi.org /10.1016/j.geomorph.2020.107284.

Yokoyama, Y., Anderson, J. B, Yamane, M., Simkins, L. M., Miyairi, Y., Yamazaki, T., Koizumi, M., et al. 2016. Widespread collapse of the Ross Ice Shelf during the late Holocene. *Proceedings of the National Academy of Science* 113:2354 –59.

Zanchettin, D., Bruni, S., Raicich, F., Lionello, P., Adloff, F., Androsov, A., Antonioli, F., et al. 2020. Review article: Sea-level rise in Venice; historic and future trends. *Natural Hazards and Earth System Sciences.* https://doi.org /10.5194/nhess-2020-351.

Zhang, P., Chen, G., Ting, M., Leung, L. R., Guan, B., and Li, L. 2023. More frequent atmospheric rivers slow the seasonal recovery of Arctic sea ice. *Nature Climate Change* 13:266–73. https://doi.org/10.1038/s41558-023 -01599-3.

Zhu, C., Langley, J. A., Ziska, L. H., Cahoon, D. R., and Megonigal, J. P. 2022. Accelerated sea-level rise is suppressing CO_2 stimulation of tidal marsh productivity: A 33-year study. *Science Advances* 8, eabn0054.

INDEX

Page numbers in italics refer to illustrations
Terms in italics refer to glossary terms

agal blooms, 193
Alabama: and Alabama Geological Survey, 173; and Alabama-Tombigbee delta, 98; and barrier islands, 229; and *bayhead deltas*, 4, 173; and beaches, 11, 90, 101, *132*, 133; and beach ridges, 130, *131*; and climate change, 173; and coastal change, 232–33; coastal plain of, 61; coast of, 7, 90, 98, 130, *131*, 133, 139, 180, 195; cypress swamps of, 178; and Dauphin Island, 11, 109, 130, *131*, *132*, 133, 139, 173, 208, 232; dunes of, 130; and East Mobile Valley, 130, *131*; erosion of, *132*, 133, 139, 229; and estuaries, 148, 232; and flooding, 173; and Florida Platform, 6; forests of, 130; Fort Gaines in, *132*, 133; and fringing wetlands, 189; Gulf Shores in, 8, 133; and Holocene deposits, 139; and lagoons, 130; Lake Shelby in, 204; and *MAFLA* (Mississippi, Alabama, and Florida) *Sheet Sand*, 11, *12*, 14, 98, 130, 139; and Mississippi Sound, 172; and Mobile and Tensaw river valley, 130, 148; and Mobile Bay, *12*, *41*, 130, 148, *151*, 153, 173–74; and Mobile bayhead delta, 173–74; and Mobile River, 173–174, 178; and Mobile Ship Channel, 133; and Mobile tidal delta, 130, 133, 173; and Mobile-Tombigbee drainage, 61; and Mobile incised valley, 173; and Morgan Peninsula, 130, *131*, 133, 173; and Pleistocene coastal deposits, 130; pollen records from, 58; polluted waters of, 178; and *progradation*, *131*; and relative sea-level rise (RSLR), *41*; and research, 232; rivers of, 138; and sand, 6, 11, 14, 61, 98, 101, 130, 133, 139, 142; and Sand Island, 130; Satsuma in, 178; and Section 309 Enhancement Grants Program, 233; and sediment, 173; and shoreline migration, 133; and storms, 130; and West Mobile Valley, 130, *131*; and wetlands, 172, 173, 189
Alcoa, 167, 178
Alley, Richard, 34
Amundsen Sea, *30*, 31
Antarctica: Amery drainage basin in, 29; and Antarctic Peninsula, 29; climate of, 58; drift of, 17; and East Antarctic Ice Sheet (EAIS), 29, 31; and glacial cycles, 21; ice discharge from, 31–32; and ice sheets, 23, 28, 29; ice streams of, 29, 31, 35; isolation of, 18; and Lake Vostok, 21; land of, 51; NASA workshops on, 51; ocean of, 51; and oxygen isotopes, 22; and Pine Island Bay, 31, 32, 51; and sea-ice, 65–66; and sea-level rise, 23, 24, 28, 29, 31, 50, 51; and seasons, 29; and sediment pile, 32; Thwaites Glacier (Doomsday Glacier) in, 162, 212; Vostok Station in, 22; and West Antarctica, 11, 28, 29, *30*, 31, 32; and West Antarctic Ice Sheet (WAIS), 29, 31, 51; and Wilkes Land-eastern Ross Sea drainage basins, 29
antecedent topography, 109
Anthropocene epoch: and coastal change, 4
Arctic amplification, 66, 69
Arctic Ocean, 65, 66
Atlantic Ocean, 198, 202–203

barrier islands, *91*; accretion of, 127; aggradation of, 120; of the Barataria-Terrebonne Estuary, 230; and Barrier Island Comprehensive Monitoring Program, 124, *125*; and beach ridges, *114*,

127, *128*, 130, 133; and Cat Island, 127, 128, *128*; and Chandeleur Island, 126, *126*, 230; and coastal change, 3, 89, 90, 104, 109, *109*, 124, 127, 128, 130, 139, vii; and Dauphin Island, 128, *129*, 130; demise of, 139; and Dog Island, 134, *135*; and Dr. Frank Stapor, 136; and East Ship Island, 127, 128; erosion of, 120, 122, 124, 128, 130, 138, 139–140, 234; and estuaries, 170–171, 212; and estuarine conditions, 127, 143; and Follets Island, 120, 122, 138, 139; formation of, 73, 90, 96, 98, 124, 126, 127, 137–138, 172; and Galveston Island, 115–118; geological research on, 127, 136–137; growth of, 104, *105*, 106, 128, 130, 136, 139; and historical time, 122, 124, 128, 133, 139, 141; and Horn Island, 127, *127*, 128, *129*; and human influences, 122; and landward migration, 120, 122, 124, 126; and Late Lafourche system, 124, 126; and longshore transport system, 130; and Louisiana's barrier islands, *125*, 138, 141, 142, 230; and Matagorda Peninsula, 122; of Mexico, *140*; migration history of, *128–129*; of Mississippi, 11, 73, 98, 104, 127–130, *127*, 138, 139, 142, 229, 232; and Mississippi Sound, *127*, 148; and Padre Island, 122; and Petit Bois Island, 127, 128, *129*; and Pleistocene ridges, 127; and progradation, 118, 127, 136–137, 138–139; restoration of, *126*; and Rio Grande, 122; and sand, *114*, 115, 120, 122, 130, 133, 136, 137–139, 141, 229, 230–231; and sea-level rise, 25, 90, 115, 130; and sediment, 106, *109*, 115, 118, 120, 122, 136, 230–231; and severe storms, *114*; and shoreline migration, 122, 141; and shore-parallel migration, 104; and South Padre Island, 120, 122; of southwest Florida coast, 136–137; and St. George Island, 134, *135*; and St. Joseph Peninsula, 134, *135*; and storm impacts, 124, 126, 130, *132*, 139, 141; and storm washover features, *105*, 106, *109*, 126, 130, 133, 139, *140*; and St. Vincent Island, 134, *135*, *136*; and Texas, 76–77, 109, 110–112, 115–118, 138–139, 141; and tidal deltas, 172; and tidal inlets, 171; and transgression, 122, 139; and upper shoreface deposits, 115; and *washover* deposits, 122; wave reworking of, 148; of west Alabama, *127*; and West and East Ship Island, *129*; and West Ship Island, 127, 128; and westward migration, 127, 128. *See also* Alabama; Louisiana; sea-level rise
bathymetry, swath, 35, *35*, 36

beach nourishment, 101–102, 142, 219, 223, 225, 227, *228*, 231, 232, 233. *See also* Florida
Bentley, Dr. Charles, 32, 34
Bindschadler, Dr. Robert, 51
Blankenship, Don, 34
Blum, Dr. Michael, 71
Brazos River, 3, 6, 10, 59, 61, 62, 73, 76, 77–78, *77. See also* Holocene epoch; Texas

Canada, 23, 40
carbon: capture of, vii; and carbonates, 6, 61, 193; and carbon dioxide, 21, 32, 185; and hydrocarbons, 162; and radiocarbon dating, 11, 56, 101, 109–110, 112, 114, 115, *116*, 118, 120, 122, 130, 136, 156, *158*, 160–161, *160*, 172, 204, 206, 207; sequestration of, 180, 186; storage of, 144
Clean Water Act (1972), 183
coastal conservation, 3, 51, 101–102, 142, 219, 227, vii–viii. *See also* beach nourishment
coastal nourishment. *See* beach nourishment
coastal sand budgets, 96, 101, 103, 120, 142
Coastal Zone Management Act (CZMA), 230
Coast in Crisis: A Global Challenge (Griggs), 200
Colorado River, 6, 10, 59, 61, 62, 63, 73, 76, *77*, 79–80. *See also* Holocene epoch; Texas
Cretaceous time, 17

data, seismic, 153, 155–156, 157, 173, 229
Davis, Dr. Richard, 225
Day, Dr. John, 71
deforming bed, 34
deltas, 160; age of, 7, 12, 71; and agriculture, 71, 84; and Alabama-Tombigbee River, 8; and Apalachicola delta, *85*, 98, 134, *135*, 138, 148, 177, 213, 216; and Apalachicola River, 8, 11, 72; and Atchafalaya delta, 73, *74*; and Atchafalaya River, 73; and *bayhead deltas*, 4, 13, 57, 59, 61, 62, 63, 64, 71, 72, 73, 76, *77*, 79, 84–88, *85–87*, 143, 152, 154–155, *154*, 155, 156–157, *158*, 160, *160*, 162, 165, 167, 168, 172–173, 176, 177, 213, 234; and Biloxi delta, 11; and "bird's foot deltas", 73; and Bolivar Peninsula, *158*; and Bolivar tidal delta, 96, 98, *158*, *159*, 162; and Brazos-Colorado delta headland, 76–78, *77*, 120; and Brazos River delta, 3, 4, *7*, 10, 11, 59, 63, 71, 72, 73, 76, 78–79, *78*, 87, 138; burial of, 60; and Calcasieu bayhead delta, 63, 177; and Caney Creek delta, 79; and *channel avulsion*, 73, 79; and chenier plains, 122–124; and clay,

123; and climate change, 4, 59, 60, 63, 76, 84; and *clinoforms*, 155–156, *155*, 160; and Colorado River delta, 4, *7*, 10, 11, 59, 71, 72, 73, 76, *80*, 87, 122, 138; and commerce, 71; and continental shelf, 59, 142; and current changes, 12; and dams, 84, 87; and deforestation, 173; and *delta headlands*, 71, 76–78, 81, 120; and delta lobes, 2, 11, 72–73, *72*, 76, 98, 124, 138, 170–171; and delta plains, 164, 165, 168; demise of, 87–88; and ebb-tidal deltas, 94, *95*, *96*, *99*, 130, *131*; and ecosystems, 84, 88; erosion of, 76, 78–79, *78*, 98, *99*, 134, 138; and Escambia delta, *86*; and estuaries, 71, 72, *158*; and flooding, 85, 88, *158*, *160*, 177; and flood-tidal delta, 96, 97, *99*, 100, *154*, *158*, 162, 166; and *fluvial-dominated deltas*, 3–4, 8, 10, 11, 70, 73, 76, 87; and Galveston Bay, 177; growth of, 11, 63, 70, 71–72, 76, 79, 79–80, 100, 118, 138, 164, 180; and historical time, 71, 75–77, 87, 98; and Holocene deltas, *40*, 70, 75–76, 180, 187; humans' impact on, 71, 79, 85, 88, 173; inundation of, 234; and Lafourche lobe, 73, 98, 124; and Lavaca bayhead delta, 63, 166; and *MAFLA* (Mississippi, Alabama, and Florida) *Sheet Sand*, 98; and Mekong delta, 45; and Mississippi River, 124; and Mississippi River Delta, 3, 10, 11, 12, 14, 43–44, 45, 64, 68, 71, 72, *72*, 73–76, *74*, *75*, 76, 87–88, 98, 122, 138, 148, 171, 174, 212, vii; and Mobile delta, 11, 61, *86*, 130, 133, 138, 173, 177, 178, 213; and Mobile River, 72; and Mobile-Tensaw bayhead delta, 190, 232, 233; of northwest Florida, *7*; and Nueces River bayhead delta, 167–168, 177, 187, 213; and offshore deltas, 11, 98, 133; and Pascaguola delta, 11; and Perdido-Escambia delta, 98, 138; and Perdido-Escambia River, 8, 11, 72; and Plaquemine delta, 73; and populations, 71; and progradation, *155*, 156, 162, 168; and Rio Grande delta, 4, *7*, 10, 11, 59, 71, 72, 73, 81–84, 87, 120, 122, *123*, 138, 212, *228*; and Rio Grande delta headland, *82*; and rivers, 72–84, 87, 88, 98; and Sabine bayhead delta, 63, 177; and Sale-Cypremort delta lobe, 73; and sand, 6, 11, 79, 83, 100, 117–118, 142; and San Luis Pass tidal delta, 96, 100, 117–118, 120; and sea-level fall, *7*, 8, 72; and sea-level rise, 4, 10, 59, 64, 71, 73, 75, 76, 84, 88, 98; and sediment, 8, 10, 62, 70–71, 73, 75–76, 78–79, 81, 84, 85, 87, 88; and *sediment isostatic adjustment (*SIA*)*,

39; and shoals, 94; and St. Bernard lobe, 73, 98, 124; and subaerial delta, 73; and Teche delta lobe, 73, 98; and tidal deltas, 11, 94, *99*, 100, 103, 117–118, 130, 133, 154, 155–156, 162; and Trinity bayhead delta, 63, 64, *87*, *158*, *159*, 160, 162, 163–164, 187, 213; and *wave-dominated deltas*, 3, 8, 71, 73, 76, *77*, *77*, 79, 81, *82*, 83, 87, 138; and Wax Lake delta, 73, *74*; of western Louisiana-Texas, *7*; and wetlands, 73, 84, 187, 190; and Yellow River delta, 45

dinoflagellates, 161
Dokka, Dr. Roy, 43
Donax, 106, *107*
dredging technology, 227, 229

Earth: and Africa, 203, 204; and agriculture, 58, 61; and Arctic changes, 69; and atmospheric rivers, 66; and *axial precession*, 19–20; and beaches, 6, 7, vii; and climate change, 1, 16–17, 19–20, *19*, 21, 22, 27–29, 32, 36, 52–58, 61, 62, 64, 69, 156, vii; and climate models, 65–66; and CO_2 (carbon dioxide), 21, 22, 32; and coastal change, 23, 52–62; and continental shelf, 71, 161; Coriolis force of, 197; and droughts, 53, 56, 66, 193; and *eccentricity cycle*, 19; and estuaries, 56, 59, 63, 71, 152–154, 161–162; and extreme weather events, 66, 69; and flooding, 42, 63, 66, 68–69, 145, 153–154, 161–162; and *fluvial incision*, 145; and *foraminifera* shells, 20, *20*; fossil pollen of, 56; and geocentric sea level, 41; and glacial cycles, 17–21, 50; and global warming, 21, 27, 28–29, 32, 50, 51, 56, *57*, 58, 65–66; and *greenhouse conditions*, 17, 21; and greenhouse gas, 27, 50, 51; and heat waves, 65, 66; and historical times, 1, 4, 7–15, 50, 56, *57*; and "hockey stick curve", 56, *57*; and human impact, 16, 25, 56, 66; and *icehouse conditions*, 17, 18, 23, 56, *57*; and ice sheets, 51; and ice volume, 20–21, 27; and interglacial conditions, 22; and Lake Agassiz, 161; lakes and lagoons of, 56; and last glacial-interglacial cycle, 4, 7–8; late Paleozoic period of, 17; and late Pleistocene, 6–9, *13*, 56, 62; *Medieval Warm Period* of, 56, *57*, 204; and megacontinent Gondwana, 17, 18; and migration, 17; and *Milankovitch cycles*, 19; and moon, 19; and North American birds, 180; and North American forebulge, *33*; Northern

Hemisphere of, 17, 18–19, 35, 37; and
  obliquity, 19; and orbit around sun, *19*;
  and oxygen isotope records, 56, *57*; and
  peninsulas, 3, 25, vii; polar climate of, 18;
  rotation of, 19, 197; and seasons, 56, 66; and
  sediment, 20–21, 56, 70–71; and Southern
  Hemisphere, 17, 66; and storm surge
  channels, 206; and temperature, 21, 22, 56,
  66, 83; and time, *19*, 22, 56; and wetlands, 8,
  56, vii. *See also* barrier islands; ice sheets;
  oceans
El Cuchillo Dam, 84
El Nino-Southern Oscillation (ENSO) cycles,
  53, 63, 202–203, 204
Emanuel, Dr. Kerry, 202
environment: and bayhead delta clinoforms,
  155–56; and carbon capture, vii; and
  chemical spills, 162; and climate change,
  4, 14, 16–17, 52–55, 59, 85, 154, 229–230,
  233–234, 235; and climate domino effect,
  69; and coastal evolution, 24; and damage
  of habitats, 185, 211; and deforestation, 162,
  173; and droughts, 83, 141; and El Nino
  and La Nina events, 53, 63; and EPA, 229;
  and estuaries, *144*, *145*, 154–155, 173–174;
  and flooding events, 156; and Florida
  Department of Environmental Protection,
  133; and foraminifera, 155; and global
  warming, *37*, 83; and greenhouse gas, 21, 27,
  225, 234, 235; and "Ike Dike Project", 225;
  and loss of ecosystems, 211; and middle-
  bay environment, 155, 156; and Mississippi
  River, 90, 141; and *ostracods*, 155; and
  pollutants, 167, 173–174, 176, 177, 178, 211;
  and rapid sea-level rise, 24, 157; and Rio
  Grande, *84*; sediment core analyses of, 156;
  and *seismic facies*, 155; and toxic wastes,
  167, 213; and tree ring records, 81, *84*; and
  undeveloped shoreline, 134; and vegetation
  patterns, 169; and wave-dominated coasts,
  90–98, 100, *107*, 108; and wetland loss,
  183–184, *194*, 211; and wildlife habitats, 180,
  183, 195
ESA-s, ERS-2, and Cryo-Sat-2 satelites, 31–32
estuaries, river-influenced, 144–148, 153–154,
  153, *154*

Falcon dam, 84
Florida, 229; and agriculture, 174; and
  Apalachicola Bay, 148, 174, *175*, 233; and
  Apalachicola Delta headland, 134; and
  Apalachicola River, 61, 133, 174; and barrier

islands, 133, 134, *135*, 136–137, 139; and
  bayhead deltas, 4; and beaches, 11, 90, 101,
  105, 133–134, 190, 231, 233; and *beach ridges*,
  105, *105*, 134, 136, *136*; Big Bend area in,
  4, 61, 185, 190–192, *191*, 193, 195, 200–201,
  212–213, 231; and Caloosahatchee and
  Peace Rivers, 136, 176; and Caloosahatchee
  River dredging, 218; and Chattahoochee
  River, 174; and climate change, 230,
  233–234; climate of, 53, 55, 58, 192, 231; and
  coastal barriers, *135*, 190; coastal plain of,
  61; coast of, 7, 55, 68, 90, 94, 98, 133, 136,
  139, 176, 180, 231; and Comprehensive
  Everglades Restoration Project (CERP),
  219; continental shelf of, 8, 12, 98, 133, *135*;
  and "critical erosion areas", 231; and cypress
  swamps, 71–72; deltas of, 8, 134, *135*, 190;
  erosion of, 133–134, 136, 139, 207, 233; and
  Escambia delta, 190; and Escambia River,
  133; and estuaries, 148, 152; and Everglades,
  4, 55, 180, 185, 193, *194*, 212, 216, 218–219,
  233, 234; and Everglades project, 218–219;
  Flint River of, 174; and flooding, 68, 201;
  and Florida Bay, 193, *194*; and Florida
  Coastal Management Program, 233; and
  Florida Department of Environmental
  Protection, 133; and Florida Keys, 231;
  and Florida Panhandle, 55, 94, 204; and
  Florida Platform, 6, 8, 39; and Florida
  State University, 195; Fort Myers in, 137;
  and freshwater, 174, 176, 192, 193, 213;
  and fringing estuaries, 190; and fringing
  wetlands, 189; and Gasparilla Island,
  136, 207; Gasparilla Sound-Charlotte
  Harbor estuary in, 176; growth of, 176;
  Hernando Beach in, 190; high tides in,
  68; and Holocene deposits, 139, 148, 192;
  and human occupation, 133, *134*, 136, 176,
  190, 212; and inlet migration, *105*; and
  Juncus roemerianus, 192; levees in, 219;
  and *MAFLA* (Mississippi, Alabama, and
  Florida) *Sheet Sand*, 11, *12*, 14, 133, *135*, 139;
  and mangrove swamps, 193; and mangrove
  wetlands, 180; and marshes, *105*, 195; and
  Miami, 215; Miracle Strip of, 133–134, 216;
  Naples in, 136; and North Captiva Island,
  136, *137*; and Okaloosa Island, *134*; oyster
  population of, 174, 192; and Panama City,
  133, 134; and Pensacola Bay/East Bay/
  Escambia Bay complex, 148; Pensacola in,
  12, 48–49, *48*; and Perdido River, 133; pollen
  records from, 58; precipitation in, 55; and

progradation, 134, 136–137; and relative sea-level rise (RSLR), *41*; rivers of, 138, 174; and salinity regimes, 174; and sand, 6, 11, 12, 61, 101, 133, 134, 136, 139, 142, 148, 233; and sand ridges, *135*; and Sanibel Island, *105*, 136–137, *137*, 207; and Santa Rosa Island, *134*; saw grass (*Cladium jamaicense*), 193; and science, 233–234; and seagrass, 192, 195, 218; and sea-level rise, 41, 43, 48–49, *48*, 55, 174, 212, 233; and sediment quality, 150, 176, 192; and shoreline migration, 133; and South Captiva Island, 136, *137*; and South Florida Water Management District, 219; southwest coast of, 136–137; and St. Marks National Wildlife Refuge, 195; and St. Marks River, 192; and St. Petersburg, 215; St. Vincent Island, *105*, 174, 207–208, *207*; and subsidence, 8, 41; and Tampa Bay, 136, 150, 176, 213, 216; Tampa in, 176, 200, 215; topography of, 193; and transgression, 133, 136; and tropical vegetation, 55; urbanization as threat to, 176; and wetlands, 176, 180, 185, 189, 193, 212, 218, 233; and wildlife, 190

fossil fuels, 21

Galveston Bay Storm Surge Barrier System, 223

geograhic information system (GIS) software, 184

glaciers: collapse of, 39; deglaciation of, 32–34; and glacial lakes, 23; ice loss from, *27*, 32; ice streams of, 31, 37; and interglacial conditions, 22; melting of, 25, 26, 27, 49; Pine Island Glacier, 31; and satellite measuring methods, 31–32; satellite obsevations of, 51; and sea-level rise, 17, 23, 25, 26, 27–28, 32, 39, 49, 50, 51, 235; Thwaites Glacier (Doomsday Glacier), 31, 32, 39, 51; and water, 23, 27, 50

Global Positioning System (GPS), 42, 44, 103

Greenland, 23, 28, 37, 49–50, 212. *See also* ice sheets

Griggs, Dr. Gary, 200

ground-penetrating radar (GPR) data, 112

Gulf Coast. *See* US Gulf Coast

Gulf Islands National Seashore, 127, *134*, 232

Gulf of Mexico, *113*, *121*; and Arctic amplification, 69; and barrier islands, 148; and chenier plains, 122; and climate change, 48, 55, 69; coastal currents of, 68–69; eastern region of, 68; and East Matagorda

Bay, *80*; ecosystem of, 180; and estuaries, 3, 143, 169, 174, 207; and flooding, 68–69, 177; and human impact on rivers, 141; and hurricanes, 198–199, 202, 203, 207, 208; industrial centers on, 213; and inlets, 164; and *Intertropical Convergence Zone* (ITCZ), 68; and lagoons, 3; Loop Current of, 47–48, *49*, 68–69, 199, 206; and Mississippi River, 81; navigation channel to, *79*; northern region of, 3, 66; and oceanographic circulation, 47–48, 66, 68–69; and Rio Grande, *82*, *83*, 84; and salinity, 94, 150, 207; and sand, 66, 68; and sea-level record, 5; and sea-level rise, 5, 10, 16, 23, 25, 43, 48, *49*, 110, 161–162; and seasonal temperatures, 68; and sediment, 59–60, 66; and transgressive events, *119*; and transgressive surface, *123*; tropical disturbances in, 63; water temperatures of, 26–27, *26*, 49, 53, 55, 68, 69, 199, 208, 230; western sector of, 43, 68, 69; and wetlands, 3, 190; and winds, 66, 68; and Yucatan Channel, 48, 68. *See also* Gulf of Mexico Basin

Gulf of Mexico Basin, 6, 39, *40*, *41*, 58–59, 62, 72

Gulf Stream, 48

Holocene epoch: and Apalachicola Bay, 161; and Brazos River valley, 156–157; and climate change, 152, 167, 168; climate of, 10, 56, 168; and coastal change, 25, 53, 110; and coastal deposits, 14, 15; and Colorado River valley, 156–157; and deltas, 87; and early Holocene, 10, 11–13, 23–24, 25, 35, 53, 58, 59, 71, 85, 98, 122, 134, 152, *153*, 156, 160, 161, 164, 166, 167, 176, 177; and estuary evolution, 156–162, 166, 180; and flooding events, 157, 167, 168, 174, 184; and global temperature, 22; and Holocene sediment, 75; and human coastal occupation, 4, 14, 210; and *ice streams*, 11, 35; and late Holocene, 10, 11, 14, 53, 56, 61, 62, 63, 70, 72, 76, *77*, 81, 85, 98, 100, 102, 110, 118, 124, 133–34, 137–39, 141, 142, *153*, *158*, 162, 164, *165*, 166, 167–69, 171, 172, 173, 177, 180, 186; and marshes, *91*; and middle Holocene, 10–11, 85; as a postglacial period, 23; and progradation, *158*; and radiocarbon dating, 11; and scorched tree rings, 56; and sea-level rise, 10, 14, 15, 157, 160, 167, 176; and sediment, 10–11, *40*, *41*, 44, 45, 60, 156–57; and seismic facies, 157; and Texas, 12–13, 60

Hughes, Dr. Terry, 31
hurricanes, *196*; and barrier islands, 232; and
    beach ridges, 206–208; and Bermuda High,
    206; and Bolivar Peninsula, 112; and the
    Caribbean, 204; categories of, 199–200,
    202, 206; and climate change, 202–203, 208;
    and coastal barriers, 227; and contaminated
    sediments, 167, 174, 177; damage from, 200,
    201, 202, 208, 212, 229; and Dauphin Island,
    132; and dinoflagellates, 207; and dust
    clouds, 203; and Florida, 204, 215, 233; and
    Follets Island, 120; formation of, 197–198;
    and Galveston Hurricane, 200; and Gulf
    Coast cities, 200, 215; Hurricane Andrew,
    200; and Hurricane Camille, 128, 201; and
    Hurricane Frederic, 208; and Hurricane
    Harvey, 201–202, 208; and Hurricane
    Helene, 177, 198, 200–201, 202, 215, 233;
    Hurricane Ian, 177, 201, 215, 233; Hurricane
    Idalia, 177; Hurricane Ike, 208, 223; and
    Hurricane Katrina, 76, 126, *126*, 128, *132*,
    172, 185, 201; and Hurricane Michael, 200;
    and Hurricane Milton, 177, 198, 201; and
    Hurricane Rita, 185; impacts of, 78, 112,
    120, 124, 126, 128, *132*, 141, 183, 185, 194, 199,
    200–202, 203, 208–209, 215, 229, 230, 234,
    vii; intensification of, 69, 199, 200–201,
    230, 233, 234; and Louisiana's barrier
    islands, 124, 126; and megahurricane
    records, 206–208; and Mississippi Sound,
    172; and Mobile Twin Hurricanes, 128;
    and National Hurricane Center's SLOSH
    model, 200; and Okeechobee Hurricane,
    200; and *paleotempestology*, 203–204, *205*;
    rapid intensification of, 201; and sand
    transportation, 100; and *storm surges*,
    199, 200, 201, 208, 223; and storm tides,
    180; and Tampa Bay, 177; and Texas, 204;
    threat of, 193; tracks of, *198*, 199, 203; and
    Velasco's destruction, 77; and vertical wind
    shear, 203; and *washover* deposits, 100; and
    washover fans, 206; and water quality, 176;
    and West and East Ship Island, 128; and
    winds, *198*, 199–200, 201, 203, 208
hypoxia, 64

ICESat (NASA), 31–32
ice sheets: and Antarctic Ice Sheet, 18, *18*, 21,
    23, 28, 29, *30*, 31–32, 34, 37, 161, 212; and
    Antarctic seafloor, 35; and Arctic seafloor,
    35; and atmospheric and oceanographic
    influences, 8, 22–23, 32–34, 37; and climate

cycles, 8, 23; and coral reefs, 23; death mask
    of, 35; decay of, 6, 8, 10, 23–24, 25, 28, 29,
    37; and East Antarctic Ice Sheet (EAIS),
    29, *30*, 31, 34; and eustatic cycles, 21; and
    glacial isotasy, 32, *33*; and greenhouse
    gas, 32; and Greenland Ice Sheet, 21, 23,
    28, 34, 37, *38*, *38*; and icebergs, 29, 31;
    and ice shelf formation, 36; and imaging
    technology, 37–39; and interglacial mode,
    21–22; and isostatic depression, 39; and
    isostatic rebound, 39–40; lakes beneath,
    34; and Laurentide Ice Sheet, 39; melting
    of, 26, 27, 28–29, 37, *38*, 49–50; and North
    American forebulge, 39, *40*; and northern
    Asia, 23; and northern Europe, 23; and
    Northern Hemisphere, 18–19, *18*, 21, 23, 37;
    and oxygen isotopes (O18 and O16), *20*;
    pressure melting of, 29; rapid changes in,
    51; and Russia, 23; and satellite measuring
    methods, 31–32, 34; and sea-level fall, 4, 6,
    8; and sea-level rise, 6, 17, 22, 22–24, 28, 31,
    32–39, 49, 50, 212, 235; and sediment, 34,
    35; and Southern Hemisphere, 18, 23; and
    subpolar latitudes, 18; and US Gulf Coast,
    39; and water, 20, 23, 28, 29, 34; and West
    Antarctic Ice Sheet (WAIS), 29, *30*, 51; and
    Wilkes Land-eastern Ross Sea drainage
    basins, *30*
ice shelves: and Filchner-Ronne Ice Shelf, *30*,
    31; and ice cliff formation, 35; and Ross Ice
    Shelf, *30*, 31
ice streams, 31, 34–35, 37
Illinois River, 76
Ingleside Paleoshoreline, 7–8, 21, 148, *149*
Intergovernmental Panel on Climate Change
    (IPCC), 2, 27, 28, 49, 50, 152
Intertropical Convergence Zone, 204, 206
Intracoastal Waterway, 185, 187, 188
IPCC *Sixth Assessment Report*, 49, 50
irrigation systems, 62

jet stream, 66, *66*, *67*, 198, 206

La Nina events, 53, 63, 81, 203
Last Glacial Maximum (LGM), 18, 23, *33*, 35,
    39
LiDAR imaging, 103, 104, *105*, *113*, *116*
Little Ice Age, 27, 42, 56, *57*
Louisiana: and Atchafalaya Bay, 170; and
    back-barrier bays, 124; and Barataria
    Bay, 170, 216–217; and Barrier Island
    Comprehensive Monitoring Program, 124,

125; and barrier islands, 124–126, 138, 141, 142; and Baton Rouge, 88; and *bayhead deltas*, 4; beaches of, 6; and Bonnet Carre Spillway, 45, 172; and Calcasieu Chenier Plain, 122; and Chandeleur Island, 124, *125*, 126; and Chandeleur Sound, 170, 172; chenier plains of, 186; climate of, 53, 55; coastal barriers of, 227; and coastal change efforts, 216–217, 230; Coastal Master Plan for, 76, 186; coastal plain of, 62; Coastal Protection and Restoration Authority of, 73; coast of, 6, 15, 51, 60, 73, 101, 122, 124, 172, 211, 213, 234; continental shelf of, 9, *9*, *12*, 59, *60*, 72, 76; and dams, 76; and Deepwater Horizon settlement, 217; and delta lobes, 72, 76, 170–171; deltas of, 7, 8, 39, 73, 75–76, 170–171; and estuaries, 124, 144–145, 170–171, 177; and flooding, 73, 76, 124, 172; forebulge collapse in, 40; and Grand Isle, 124, *125*; and Holocene strata, 14, 45, 75–76; and human impact, 25, 172; and jetties, 124; Lake Borgne in, 172; and Lake Pontchartrain, 172; land loss in, 4, 45, *46*, 49, 73, 76; Layfayette in, 88; levee system in, 76; and Louisiana Chenier Plain, 4, 45, 90, *91*, 212, vii; and Mississippi River, 124, 141, 185; and Mississippi Sound, 171; Morgan City in, 73; and Natural Resource Damage Assesment, 217; New Orleans in, *40*, 45, 62, 76, *125*, 200, 201, 215; ports of, 62; precipitation in, 55; and progradation, 124; and relative sea-level rise (RSLR), *41*, 45; and research, 234; and river-influenced estuaries, 144–145; and Sabine Chenier Plain, 122, 124; and salinity, 171, 172; and sand, 62, *91*, 101, 122, 124, 126, 142; seafood industry of, 171, 172; and sea-level rise, 1, 41, 45, 49, 51, *167*, 171, 211, 215; sediment supply in, 51, 60, 75, 124; and shoreline change, 124, *125*, 142, 227; south part of, *171*; and storms, 198; and subsidence, 17, 25, *40*, 43, 44, 45, 49; Terrebone Basin in, 44, *44*, 45; and Terrebonne Bay, 170; and Timbalier Islands, 124, *125*; and Vermillion Bay, 170; and vertical accretion rates, 186; and wave-dominated shoreline, 90; and wetlands, 25, 51, 124, 171, 180, 184, 185–186, 230; and winds, 172. *See also* deltas
Louisiana Sand Resources Database (LASARD), 231
Louisiana State University, 71
Lower Colorado River Authority, 79–80

Mann, Dr. Michael, 56
Marine Isotope Stages (MIS), 5, 6, 7, 8–9, 10
Marte R. Gomez Dam, 84
McMaster University, 134
megadroughts, 81
Mercer, Dr. John, 32
Mexico, 53, 55, 84, 139, *140*
Mexico's Laguna Madre, 139
Mexico's Tamaulipan coastline, 139
Milankovitch's astrnomical cycles, *19*, 21
Mississippi: barrier islands of, 11, 127–130, 138, 139, 142, 232; and beaches, 101; and Biloxi valley, 148; and climate change, 61, 62, 232; and coastal marshes, 185; coastal plain of, 61; and coastal plan updates, 232; coast of, 7, 180; continental shelf of, 9, *9*, 59, 142; and cypress swamps, 88; and delta lobes, 72; delta of, 11; and fringing wetlands, 189–190; and *MAFLA* (Mississippi, Alabama, and Florida) *Sheet Sand*, 11, *12*; and Mississippi Sound, 127, 142, 148; and Pascaguola River, 61; and Pascaguola valley, 148; and Pearl River, 61; and relative sea-level rise (RSLR), *41*; and research, 14, 127; and sand, 101, 142, 232; and sand budgets, 232; and University of Southern Mississippi, 14, 127, 232; and wave-dominated shoreline, 90; and wetlands, 189–190, 232. *See also* Mississippi River
Mississippi-Atchafalaya watershed, 58–59
Mississippi River, 3, 6, 11, 59–60, *60*, 61, 72–73, 76, 166. *See also* deltas; environment; Gulf of Mexico; Louisiana; rivers' drainage basins; US Gulf Coast
Morton, Dr. Robert, 128

NASA-German Aerospace Center (DLR) Gravity Recovery and Climate Experiment (GRACE) satellites, 42
National Academies of Sciences, Engineering, and Medicine, 219
National Aeronautics and Space Administration (NASA): and Gulf issues, 229; ICESat of, 42; and Jet Propulsion Laboratory, 45. *See also* Antarctica
National Fish and Wildlife Foundation's Gulf Environmental Benefit Fund, 217
National Oceanographic and Atmospheric Administration (NOAA): Coastal Change Analysis Program (C-CAP) of, 183; Coastal Inundation Dashboard of, 183, 184, 186, 187; data from, *24*, 48–49; and Gulf

issues, 229; and high-tide flooding, 69; inundation model of, 193; and LiDAR data, *101*; model results from, 2, *164*, *168*, *171*, *175*; and Northern Hemisphere ice sheets, *18*; and Office for Coastal Management (OCM), 229, 230; and prediction for Satsuma, Alabama, 178; and predictions for hurricanes, 208; and predictions for sea-level, 231, 232; and seafloor images, *35*; and sea-level rise, 2, 163, *164*, 184, *194*, *226*, 232; *Sea Level Rise Technical Report* of, 48–49; and swath bathymetry, *35*
New Mexico, 61, 62
North American Free Trade Agreement, 84

oceans: and *albedo effect*, 65; and beaches, 93; and Bureau of Ocean Energy Management (BOEM), 229; Caribbean Sea, 68; and climate change, 21, 25, 60, 65–66, 68, 69; and coastal barriers, 94, *95*; and continental shelf, 66; and corrugations, *36*, 37; currents of, 94, 95, 96; and estuaries, 95; and fair-weather waves (swell), 92, 93, *93*; and glacial water, 23–25; and global warming, 235; and *grounding zone wedges*, 35, *36*, 37; and headlands, 94, *95*; and *iceberg furrows*, 37; and *longshore current*, 94, *95*; and *longshore transport*, 94, *95*; mapping of, 42; and oceanographic circulation, 66, 68–69; and oxygen isotopes (O18 and O16), 20, *20*; and *rip currents*, 94, *95*; and sand transportation, 94, *95*; and sea-ice, 65–66; and sea-level rise, 25–28, 35–37; and sediment, 35; and shoals, *95*; storm waves in, 93; and surf zones, 93; and swash zone, 93, *93*; and temperature, 21, 25, 26–27, 32, 50, 52, 202–203, 206; and tides, 46–48; and time, 21; and tropical water temperatures, 202; volume of water in, 20, 26; warming of, 25, 26, 27, 50, 53; and wave geometry, 92, 93; and wave motion, 92–94, *95*; and wave refraction, 94, *95*; and wave's shoreface profile, 93; and wave's *shoreface zone*, 93; and winds, 66, 68, 92, 93, *95*. *See also* Arctic Ocean; Atlantic Ocean; Pacific Ocean; sea-level rise; swath bathymetry; tides
oil and gas: and Deepwater Horizon oil spill, 172, 185, 217, 230; extraction of, 44, 45, 46; and land loss, 45, *46*; and oil spills, 172, 183, 185, 194, 211, 215, 229, 230–31, 235; and refineries' locations, 215

optically stimulated luminescence (OSL), 109–110, 122, 130, 134, *136*
Otvos, Dr. Ervin, 14, 127, 128

Pacific Ocean, 53, 63
Paleocene-Eocene Thermal Maximum (PETM), 17–18
paleoclimate records, 55–58, 83
plate tectonics, 17, 50
Pleistocene epoch, 6–9, *13*, 56, 62, 71, *91*, 100, 108, 134, 148

Rangia shells, 154–155, *154*
recessional moraines, 35, 36, *36*
Rice University, 152, 155, 173
Rignot, Dr. Eric, 28
Rio Grande, 4, 6, 7, 10, 59, 61, 62, 63, 73, 76. *See also* deltas; Texas
rivers' drainage basins, 58–59, *59*, 61–62, 63, 73, 81, 162, 166, 172, 173. *See also* Florida
Roberts, Dr. Harry, 71
Ross Sea, 30, 35

Saffir, Herbert, 199
Saffir-Simpson Hurricane Wind Scale, 199–200
Saving America's Amazon (Raines), 173
sea-level rise: acceleration of, 225, 226–227, 230; and atmospheric temperature, 22; and *back-stepping*, 157; and barrier islands, 25, 90, 130, 141; and "bathtub" models, 183–184; causes of, 22–29, 31, 35–39; changes in, 3, 5, 6, 10, 23–24, *24*; and city of Freeport, 78; and climate change, 232, 233, 234, 235; and coastal barriers, 13, *13*, 102–103, *102*, 111, 122, 139; and coastal change, 183, 232, 233–234, *234*; and coastal deposits, 23; and coastal evolution, 141, 142; coastal response to, 42, 51, 89, 92, 102, 141; and Coastal Vulnerability Index (CVI), 213; and deltas, 4, 59, 63, 64, 71, 88, 98, 143, 190; and early Holocene period, 10, 11–13, 23–24, 25, 35, 166; and ecosystems, 84, 88; and estuaries, 11, 85, 144–145, *149*, 152, 153–154, *153*, 160–161, 166; and eustatic rise (*eustacy*), 17, 24–25, 73; and federal agencies, 229; and flooding, 153–154, 156, 161–162, 215; and Florida, 48–49, 55, *175*, 219, 233; in the future, 162, *167*, 168, *171*, *175*; and *geodetic benchmarks*, 40–41, 43; and glacial cycles, 17–25, 42, 50; and global cooling, 42; and global warming, *50*, 51; and greenhouse

gas, 17, 22–23, 22, *50*, 51, 234; and grounding line retreats, 35–37; and groundwater, 28; and Gulf Coast, 2, 101; and human influence, 22, 28; and human influences, 211–212; and hurricanes, 199, 209; and ice sheets, 22–24, 27, 28–29, 31, 32–39, 235; and interglacial conditions, 22; and inundation, 166, 168, 171, *171*, 174, *175*, 235; and IPCC *Sixth Assessment Report*, 28; and land loss, 4, 25, 49; and landward migration, 174, 189; and late Holocene, 10, 141, 171; and *lowstand* period, 153; and Marine Isotope Stages (MIS1-6), 5; and *mean rate of sea-level rise* (MSLR), 41; measurement of, 40–42; and middle Holocene, 10; and Mobile Bay, 41; and NOAA's models, *167*, 168, *168*; and North Carolina's coast, 42–43; and Nueces Bay, 168; and numerial models, 102–103; and oceanographic influences, 48–49; and ocean thermal expansion, 26; and offshore Sabine Valley, 157; and offshore Trinity Valley, 157; and peninsulas, 25; and punctuated sea-level rise, 5, 13, 23, 35, 39, 156, 157; and rapid sea-level rise, 24, 157, 161, 174, 210, 230; rate of, 2, 10, 11, 14, 16–17, 22–23, 24, 25, 28, 37, 41, 42–43, 48–49, 53, 59, 64, 71, 72, 73, 76, 85, 104, 110, 115, 137, 152, *153*, 162, 168, 177, 211–212; records of, 21, 23, 24, 40–42, 42–43, *43*, 176–177; and relative sea-level rise (RSLR), 17, 41, *41*, 43, 44, 45, 70, 73, 75, 76, 84, 165, 168, 184, 186, 193, 208, 213, 215, 231; research on, 152–153; response to, 184–185; and sand supply, 25, 104; and satellite measuring methods, 41–42; and seagrass meadows, 179–180; and Sea Level Affecting Marshes Model (SLAMM), 184; and sediment supply, 61, 64, 184; and shoals, 127; and shoreline migration, 48, 103, 104; and South Carolina's coast, 42–43; and *steric influence*, 25, 26–27; and subsidence, 39, 41; threat of, 211; and tides, 46–48; and time, 5, 17, 24, 49, 51; and vertical accretion rates, 184, 186; and water heating, 48, 49; and wetlands, 90, 152, 183–184, 185, 187, 194, 208. *See also* Antarctica; glaciers; Gulf of Mexico; subsidence; tide-gauge records

Shinkle. Dr. Kurt, 43

Simpson, Robert, 199–200

subsidence: basin subsidence, 39, *40*; causes of, 44–50; and chenier plains, 186; and continental shelf, 8, 72, 98; definition of, 39; and eustatic sea-level rise, 49; and faulting, 39, *40*, 41, 44; in Florida, 8, 192; and *glacial isostatic adjustment* (GIA), 39, 40, *40*; and glacial isostatic processes, 50; and groundwater, 45, 46, 47; histories of, 6, 39; and Holocene sediments, *40*, 44–45, 177; and Houston Industial Complex, *47*; and human activity, 45; and land loss, 50–51; and Louisiana's sites, 231; measurements of, *44*, 45; and Mississippi River Delta, 186; and oil and gas, 211; rates of, 17, 39, 40, *40*, 43, 44–45, 72, 73, 75, 98, 124, 141, 166; and relative sea-level rise (RSLR), *41*, 42; and salt migration, 39; and sediment compaction, 39, 50, 75; and *sediment isostatic adjustment (*SIA), 39, *40*; and subsurface fluid extraction, 45; and subsurface salt migration, 39; and technology, 231; and tectonic movement, 50; and *tectonic subsidence*, 39, *40*; and Texas, 231; and US Gulf Coast, 1, 6, 39, *40*; variations of, 42; and water production, *46*; and wetlands, 184. *See also* deltas; sea-level rise

Swamp Land Acts, 180

Texas: and 2021 Texas Freeze, 66; and aggradation, *121*; and agriculture, 162, 166; and Aransas Bay, 166; Arctic air in, 66; aridity in, 55, 57, 58, 62, 63, 73, 76, 81, 87, 98, 143, 150, 152, 169, 170, 177, 187; and Baffin Bay, 56–57, 148, 150, 153, 166, 169; and *bayhead deltas*, 4, 57, 63–64; and beaches, 99, 112, *113*, 115–116, 120, 142, 223–225; and beach ridges, 112, 115, *116*, 118; and Bolivar Peninsula, *99*, 112, 113–115, 118, 130, 139, 164, 206, 219, 223; and Bolivar tidal inlet, 112, *113*, 115; and Brazos River, 63, 64, 76, 77–79, *77*, 87, 138, 141; and Calcasieu River, 166; and Caney Creek, *77*, 79, *80*; Central coast of, 118–119; and Christmas Bay, 120, *121*; Clear Lake area of, 46; and climate, 1, 53, 56–58, 62, 64, 76, 110, 122, 152, 162, 169, 187; and climate change, 231; and climate models, 152; and coastal barriers, 68, 76, *101*, 110–12, 118, 138, 141, 231; and coastal change, *111*, 118–119, 231; and coastal composition, 6, 8, 12–13, 15, 60, 62, 76–77, 80; and Coastal Erosion Planning and Response Act (CEPRA), 231; and coastal marshes, 185; and *Coastal Texas Protection and Restoration Feasibility Study*, 231, 232;

coastal waters of, 60; coast of, 62, 64, 94, 98, 100, 102, 110–112, 118–119, 138, 142, 152, 166–170, *167*, 169, *170*, 213; and Colorado River, 64, *77*, 79–81, 138; and Commision on Environmental Quality, 230; and continental shelf, 9, *9*, 12, *13*, 14, 59, 60, 72, 98, 100, 122, 142; and Copano Bay, 166; and coral reefs, 60; and Corpus Christi Bay, 57–58, 64, 153, 166, 167, 168, 169, 213; and cypress swamps, 71–72, 88; and dams, 79, 80, 169; deltas of, *7*, *7*, 8, 39, 57, 63–64, 76–85, 138, 166; and dredge spoil banks, 188; dune fields of, 55; and East Matagorda Bay, *80*; and Enhanced Dune and Beach System, 223, 225, 231; erosion of, 76–77, 79, *82*, 100, 102, 112, 114–116, *114*, 117, 118, 120, 139, 142, 168, 188, 206, 219; and estuaries, 45, 58, 63, *85–87*, *114*, 120, 144–145, 148, 152, *158*, 162–170, 177; and flooding, 85, *158*, 202, 208, 209; and Follets Island, 76, *77* 96, 109, 120, *121*, 138, 139; forests of, 55, 166; and fossils, 114; and Freeport Channel, *78*; Freeport in, 78, 79, 213, *214*; and freshwater, 143, 162, 166–167, 169; and Galveston Bay, 58, 63, 64, 112, *113*, 114, 148, 150, 153, *158*, 162–163, *163*, 168, 174, 186, 213, 231; and Galveston Beach, *221*; Galveston in, 40, 45, *47*, 77, 98, 223; and Galveston Island, *95*, 96, 98, *99*, 100, *105*, 110, 115–18, *116*, 120, 139, 187, 208, 209, 219, 223; and Galveston Seawall, 98, 115, *116*, 117, 219, *222*, 223; General Land Office of, 102, 223; and Great Storm of 1900, 115, 219; and Herald Bank, *13*; and historical shoreline change, 111–12, *111*, 120; and historical time, 114, 120; and Holocene sediments, 45, 110; and Houston/ Galveston, 200; and Houston/Galveston Ship Channel, 46, 98, *99*, 115, 162–163, *163*, *220*; Houston in, *47*, 178, 201, 202, 208, 213, 215; humans' impact on, 162–163; humidity in, 63; and hurricanes, 206, 208; and Ike Dike Plan, 223–225, 227, 229; industrial centers in, 78, 162, *163*; and jetties, *99*, 115, *220*, *221*; and lagoons, 169, 170; and Laguna Madre, 122, *123*, 139, 150, 166, 169, *170*, 180; lakes of, 64; land loss in, 166; and landward migration, 120, 122, 169, 231; and late Pleistocene, *13*; and Lavaca Bay, 167, 174; and Lavaca River, 167; levees of, *214*; and lock construction, 163; logging in, 166; and longshore transport system, 98, *99*; and marsh deposits, *114*; and Matagorda Bay,

153, 166, 167, 188, *188*, 213; and Matagorda Bay Ship Channel, 167; and Matagorda Island, 118; and Matagorda Peninsula, 76, *77*, 79, 81, *101*, 109, 120, 138, 139, 188, *188*; and Mustang Island, 118, *119*, 161; and Nueces Bay, 161; and Nueces River, 63, 169; and Nueces River bayhead delta, 57–58, 64; oyster reefs of, 57–58, 161, 163, 166, 169; and Padre Island, 57, 64, 110, 122, *123*, 169–170; and Point Comfort aluminum processing plant, 167; and pollutants, 162, 167; and Port Arthur, 215; and Port Mansfield, 169, 170; power grid of, 65; and precipitation, 53, 56, 58, 59, 63, 87; and progradation, 112, *113*, 114–115, *116*, 118, *119*; and relative sea-level rise (RSLR), *41*; and Rice University drill cores, *158*; and Rio Grande, 64, 87, 138, 141, 148; and river-influenced estuaries, 144–145; rivers in, 62–64, 98, 138, 141; and Sabine Bank, *13*; and Sabine River, *13*, 63, 166; and saline conditions, 152, 161, 163, 166, 169, 170, 177; and San Antonio Bay, 166; and sand, 11, 13, 62, 64, 76, *77*, 80, 98, *99*, 100, 101–102, 112, *113*, *114*, 115, 117–118, 120, 122, 142, 219, 223, 225, 227, 231; and San Luis Pass, 120, *121*; seagrass of, 169; and sea-level rise, 1, 13, 48, *167*, 168, 170, 215; and sediment, 8, 12–13, 60, 61, 62, 64, 73, *77*, 80, 98, 112, 114–115, *116*, 118, *119*, *121*, 141, 161, 166, 169, 170; and seismic profiles, 112; and Shepard Bank, *13*; shoreface erosion in, 100; and shoreline change, 227; and shoreline migration, 112, 117, 138; and South Padre Island, 109, 120, *123*, 138, 139; and St. Joseph Island, 118; and storms, 198; and subsidence rates, 45, 46, *47*; and Surfside Beach, 120; and Tampa Bay, 65, *152*, 161; and temperature, 56, 65; and Texas Mud Blanket, 12, 60, 81, 138, 227, *228*; and town of Velasco, *77–78*; and transgression, *121*, 139; transgressive barriers of, 120–122; and Trinity delta, 163–64; and Trinity River, *13*, 63, 162; and Trinity Valley, 162; and tropical dry forests, 55; and *washover* deposits, *101*, 120; and *washover fans*, *101*, 112, *113*; and wave-dominated shoreline, 90; and West Bay study, 187; and wetlands, 166, 180, 184, 186–88

Texas Bureau of Economic Geology, 103
Thomas, Dr. Robert, 32, 34
tide-gauge records, 24, 40–42, *43*, 45, 46, *47*, *48*, 168

tides: *Astronomical tides*, 46, 47, *48*, 68; and beaches, 95; circulation of, 150, *151*, 152, 166; and ebb-tidal deltas, 94, *95*, *96*; and ebb tides, 96, 150, *151*; and estuaries, 150; and flood-tidal delta, *95*, *96*; and flood tides, 96, 150, *151*; and gauges in estuaries, 231; and inlets, 166; and "king tides", 68; *Meterological tides*, 46, 47, *48*, 68; and microtidal settings, 92; and oceanographic circulation, 66; *Oceanographic tides*, 47, *48*, 68, 69; and sediment delivery, 150; and *Spartina alterniflora*, 180; and storm tides, 180; and tidal forces, 19–20, 96; and tidal inlets, 94, *95*, 133, 150, 155–56; and tide-dominated coasts, 190. *See also* tide-gauge records

Tornqvist, Dr. Torbjorn, 44–45

transgression, 10–12, 100, 104, 106, 108, 109, *109*, 112, 114–15, *116*. *See also* barrier islands; Florida; Texas

tropical storms, 197, 199, 201, 202, 204. *See also* hurricanes

Tulane University, 166, 186

Union of Concerned Scientists, 64–65

United States: Atlantic Coast of, *24*, *43*; atmospheric temperatures in, 53; and climate models, 62, 63–64; Congress of, 127; Corps of Engineers of, 167; and development of coasts, 133; drainage systems in, 172; droughts in, 53, 141; economy of, 211, 233, 235; and estuaries, 183; federal agencies of, 229; Georgia in, 174; and Government Accountability Office, 215; heat waves in, 53; ice sheets on, 23, *33*, 39; and Ike Dike Plan, 227; and Intergovernmental Panel on Climate Change (IPCC), 2; and jet stream, *66*, *67*; and Laurentide Ice Sheet, 39; Massachusetts in, 184; and National Wetlands Inventory, 183; and NOAA's *Sea Level Rise Technical Report*, 49; and polluted waters, 173–174, 183; precipitation in, 53, 63, 81, 83, *84*; and sand, 63; southwestern part of, 53, 62, 63, 81, 83, 169; and subsidence, 40; Superfund sites in, 167; and US Army Corps of Engineers, 73, 141, 219, 223, 225; use of tide-gauge records in, 40–41; warm conditions in, 66; and Water Resources Development Act (2000), 219; and wetland loss, 183. *See also* Alabama; Florida; Louisiana; New Mexico; Texas; US

Gulf Coast

United States Geological Survey (USGS), 103, 110, 133, 155, 173, 174

University of Alabama, 152, 155, 173

University of California (Santa Barbara), 152–153

University of California (Santa Cruz), 200

US Geological Survey (USGS) Coastal Change Hazards website, 213

US Gulf Coast: and *aggradation*, 104; and agriculture, 64, 65, 180, 183; and Alabama-Tombigbee River, 6, 59; and ancillary basins, 61; and Apalachicola River, 6, 59; and aquatic life, 66; aridity in, 83; and *Artic amplification*, 66; and atmospheric and water temperatures, 52–53, *54*, 66; and "bathtub" models, 3, 102; and *bayline*, 106, *107*; and *beach ridges*, 104, *105*, 106; and Bonnet Carre Spillway, 212; and brackish marshes, 180; and Brazos River basin, 64; and breaker zones, 108; and Calcasieu chenier plain, 137; and Calcasieu Lake, 122, 148, 150, 153, 164–166, *165*, 177; and chenier plains, 4, 45, 90, *91*, *92*, 94, 98, 100, 110, 122, 124, 137–138, 154, 227, 234, vii; and chenier ridges, *91*; and climate change, 1, 2, 52–53, 55, 61, 62, 65, 88, 92, 101, 103–104, 144, 177, 210–211, 229–230, 233–235, vii–viii; climate zones of, 53; and coastal barriers, 90, 96, 100, *101*, 102–103, 104, 105–106, 109, 110, 154, 162, 227; and coastal change, 2, 3–4, 10–11, 14, *46*, 49, 51, 55, 71, 90, 100, 103–104, 106, 109–112, 141, 210–11, 212, 229, 230–31, 233–35, vii, viii; coastal hazards map for, *215*; and coastal lakes, 90; and *coastal plain* topography, 61; and continental shelf, 138; and Copano Bay, 153; cypress swamps on, 71–72, 88; and dams, 58, 64, 76, 79, 80, 84, 87, 122, 141, 162, 169, 173, 177; and dead zones, 64–65; decline of, 216; and deglaciation phases, 10; DeSoto Canyon in, 6; and droughts, 65, 66, 81, 83, *84*, 87; eastern sector of, 6, 53, 58; erosion of, 25, 45, 49, 55, 58, 70, 72, 83, 90, 98, 100, 109, 138, 139, 141–142, 184, 226, 227, 229, 233; and estuaries, 3, 8, 45, 58, 64, 65, 88, 90, 94, 96, 106, *107*, 108, 143–178, *145*, *146*, *147*, 170–78, 187, 190, 212, 213, 230, 231, vii; and evolution of coast, 14–15, 25, 39–40, 49; and extreme weather events, 65–66; and floods, 65, 66, 148, 153–154, 160–161, *165*, 176; and fluvial terraces, 185;

forests of, 55, 58; and freshwater, 58, 143, 144, 150, 154, 162, 169, 171–72, 180; and freshwater marshes, 180; and fringing wetlands, 194; and Galveston Bay, 96, 97, 162–164, 177, 207, 213; and hazardous chemicals, 215; and historical time, 62, 103–104, 110, 111–112; and Holocene sediments, 45; and human influences, 2, 3, 14, 45, 46, 58, 62, 79, 85, 100, 110, 112, 141, 162, 172, 177, 178, 180, 225–226; and hurricanes, 206–209, 234; industrial centers on, 46, 78, 85, 162, 194, 208, 234, vii; *Ingleside Paleoshoreline* of, 7–8, 21, 148, 149; and inlet migration, 105, 106; jetties on, 98, 141, 219; and *Juncus roemerianus*, 180, 182; and lagoons, 8, 90, 96, 106, 148, 169, 212; and Laguna Madre, 4, 139, 169, 170; and land loss, 49; and landward migration, 142, 148, 149; and large rivers, 59, 96, 141; and late Pleistocene, 98; and *longshore transport*, 96; and *longshore transport*, 98; and *longshore transport*, 100, 103; and Louisiana Chenier Plain, 4, 45, 166, 212, vii; and lower shoreface zones, 108; and *MAFLA (Mississippi, Alabama, and Florida) Sheet Sand*, 14, 138, 229; and mainland beaches, 7, 11, 90, 93; and Matagorda Peninsula, 120, 122; microtidal setting of, 92; and Mississippi Delta plain, 88; and Mississippi Gulf Coast, 127–130, 212; Mississippi River drainage basin of, 62; and Mississippi Sound, 171–172, 190, 212; and Mobile Bay, 96, 96, 150, 151, 161, 171, 172–174, 175, 177, 190, 213; and Mobile River, 173–174; modern topography of, 6; and oak-dominated forests, 58; and oxygen isotopes, 22; and Padre Island, 122, 169–170; and Pascaguola Bay, 171; and peninsulas, 89, 90, 91, 96, 106, 109, 112, 120, 130, 131, 134, 138, 139, 143, 212; and pine-dominated forests, 58; Pleistocene shorelines of, 185; and populations, vii; and precipitation, 52, 53–55, 56, 58, 61, 63, 83–84, 144, 152; and *progradation*, 104, 105, 106, 109, 110, 122, 139, 142; records of, 103–104, 142; and rivers, 6, 58–62, 71, 72, 84, 90, 110, 172; and runnels, 108; and Sabine Chenier Plain, 91, 92, 122, 137; and Sabine Lake, 91, 92,

96, 97, 122, 148, 150, 153, 164–166, 177; and salinity, 144, 150, 164, 171, 177, 180; and saltmarsh vegetation, 180; and sand, 12, 13, 14, 90, 91, 92, 94, 96, 98, 99–100, 101, 103, 104, 106, 107, 108, 110, 137–138, 226, 227, 229, 231, 233; and sandbars (ridges), 108, 109; and science, 102, 227, 230, 231, 233, 234–235, vii–viii; and seagrass meadows, 179–180, 182; and sea-level fall, 8, 71, 144–145; and sea-level rise, 2, 3, 4, 10, 17, 22, 24, 25, 28, 51, 52, 78, 84, 92, 101, 102, 103–104, 141, 142, 144, 183, 210, 211, 225, vii; and sediment, 3, 4, 6, 10–11, 12, 25, 45, 52, 55, 58–62, 63, 90, 92, 98, 101, 104, 107, 108, 109, 110, 123, 141, 142, 148, 150, 151, 152, 153, 162, 165–166, 172–173, 177, 179, 183, 212, 225–226; and *sedimentary facies*, 106, 107, 108, 109, 154; and sediment discharge, 59, 152; and seismic facies, 160; and shoreline migration, 103, 104, 106, 108, 109, 109, 110, 141, 142; and *Spartina alterniflora*, 180, 182; and *Spartina patens*, 180; states of, 7, 41, 229, 230, vii; and storm beaches, 106; and St. Vincent Island, 174; and subsidence, 1, 6, 17, 40–41, 43, 44–46; and swash zone, 106, 107, 108; and temperatures, 65, 66; and tidal inlets, 94–96; and tourists, 133, 144; and transgressive ravinement, 100, 108, 148; and upper shoreface zones, 108; US Geological Survey (USGS), 229; and US population, 211; vegetation zones of, 55, 58, 66, 180, 181; and *vertical accretion*, 25; and *washover fans*, 100, 106; and water temperatures, 141; and wave-dominated coasts, 142; western sector of, 6, 41, 59, 66; and wetlands, 8, 25, 61, 88, 90, 148, 152, 171, 174, 177, 179–84, 181, 185, 186–192, 193, 194, 208, 211, 212, 226, 230. *See also* Alabama; barrier islands; coastal sand budgets; deltas; Florida; Gulf of Mexico; Intracoastal Waterway; Louisiana; Mississippi; sea-level rise; Texas; transgression

Venice, Italy, 42

Wahl, Eugene, 81
Wallace, Dr. Davin, 14, 127
Wesley E. Seale Dam, 169

Other Books in the Harte Research Institute for Gulf of Mexico Studies Series

Coral Reefs of the Southern Gulf of Mexico
Edited by John W. Tunnell, Ernesto A. Chávez, and Kim Withers

Arrecifes Coralinos del sur del Golfo de México
Edited by John W. Tunnell, Ernesto A. Chávez, and Kim Withers

The Gulf of Mexico Origin, Waters, and Biota: Volume I, Biodiversity
Edited by Darryl L. Felder and David K. Camp

The Gulf of Mexico Origin, Waters, and Biota: Volume 2, Ocean and Coastal Economy
Edited by James C. Cato

The Gulf of Mexico Origin, Waters, and Biota: Volume 3, Geology
Edited by Noreen A. Buster and Charles W. Holmes

The Gulf of Mexico Origin, Waters, and Biota: Volume 4, Ecosystem-Based Management
Edited by John W. Day and Alejandro Yáñez-Arancibia

The Gulf of Mexico Origin, Waters, and Biota: Volume 5, Chemical Oceanography
Edited by Thomas S. Bianchi

Encyclopedia of Texas Seashells: Identification, Ecology, Distribution, and History
John W. Tunnell, Jean Andrews, Noe C. Barrera, and Fabio Moretzsohn

Sea-Level Change in the Gulf of Mexico
Richard A. Davis

Marine Plants of the Texas Coast
Roy L. Lehman

Beaches of the Gulf Coast
Richard A. Davis

Texas Seashells: A Field Guide
John W. Tunnell, Noe C. Barrera, and Fabio Moretzsohn

Benthic Foraminifera of the Gulf of Mexico: Distribution, Ecology, Paleoecology
C. Wylie Poag

The American Sea: A Natural History of the Gulf of Mexico
Rezneat Milton Darnell

Birdlife of the Gulf of Mexico
Joanna Burger

It's More Than Fishing: The Art of Texas Trout and Redfish Angling
Patrick D. Murray

Sea Change: A Message of the Oceans
Sylvia Alice Earle